FUNDAMENTALS OF SOILS

Soils are an essential and dynamic part of the environmental system, existing at the interface of the geosphere, hydrosphere, biosphere and atmosphere. *Fundamentals of Soils* presents a comprehensive and engaging introduction to soils and the workings of soil systems.

Starting with an introduction to the nature of soils, what soil is and the basic character of soil systems, the author moves on to look at:

- soil properties
- soil processes
- controls on soil formation
- soil classification
- world soils
- soil patterns
- soil degradation

Written in a lively and accessible manner and including a summary, essay questions and guides for further reading at the end of each chapter, *Fundamentals of Soils* provides an appealing introduction for students to all the key areas of this essential aspect of physical geography.

John Gerrard is a Reader in the Department of Geography at the University of Birmingham

ROUTLEDGE FUNDAMENTALS OF PHYSICAL GEOGRAPHY SERIES

Series Editor: John Gerrard

This new series of focused, introductory textbooks presents comprehensive, up-to-date introductions to the fundamental concepts, natural processes and human/environmental impacts within each of the core physical geography sub-disciplines. Uniformly designed, each volume includes student-friendly features: plentiful illustrations, boxed case studies, key concepts and summaries, further reading guides and a glossary.

Already published:
Fundamentals of Biogeography
Richard John Huggett

Fundamentals of Soils
John Gerrard

Forthcoming:
Fundamentals of Hydrology
Tim Davie
(October 2001)

Fundamentals of Geomorphology
Richard John Huggett
(March 2003)

FUNDAMENTALS OF SOILS

John Gerrard

Routledge Fundamentals of Physical Geography

Routledge
Taylor & Francis Group

LONDON AND NEW YORK

First published 2000
by Routledge
2 Park Square, Milton Park, Abingdon, Oxon, OX14 4RN

Simultaneously published in the USA and Canada
by Routledge
711 Third Avenue, New York, NY 10017

Routledge is an imprint of the Taylor & Francis Group

Transferred to Digital Printing 2006

Typeset in Garamond by Keystroke, Jacaranda Lodge, Wolverhampton

British Library Cataloguing in Publication Data
A catalogue record for this book is available from the British Library

Library of Congress Cataloging in Publication Data
Gerrard, John
Fundamentals of soils / John Gerrard.
p.   cm. – (Routledge fundamentals of physical geography series)
(acid free paper)
Includes bibliographical references and index.
1. Soils. I. Title. II. Series.
S591 .G44 2000
631.4–dc21   99-054629

ISBN 0–415–17004–4 (hbk)
ISBN 0–415–17005–2 (pbk)

Publisher's Note
The publisher has gone to great lengths to ensure the quality of this reprint
but points out that some imperfections in the original may be apparent

CONTENTS

Series editor's preface   vii
List of plates   ix
List of figures   x
List of tables   xiii
List of boxes   xv
Acknowledgements   xvi

1 THE NATURE OF SOIL   1

2 SOIL PROPERTIES   22

3 SOIL PROCESSES   45

4 CONTROLS ON SOIL FORMATION   72

5 HORIZONS AND SOIL CLASSIFICATIONS   101

6 MAIN WORLD SOILS   130

7 SOIL PATTERNS   152

8 SOIL DEGRADATION   177

Glossary   201
References   207
Index   225

SERIES EDITOR'S PREFACE

We are presently living in a time of unparalleled change and when concern for the environment has never been greater. Global warming and climate change, possible rising sea levels, deforestation, desertification and widespread soil erosion are just some of the issues of current concern. Although it is the role of human activity in such issues that is of most concern, this activity affects the operation of the natural processes that occur within the physical environment. Most of these processes and their effects are taught and researched within the academic discipline of physical geography. A knowledge and understanding of physical geography, and all it entails, is vitally important.

It is the aim of this *Fundamentals of Physical Geography Series* to provide, in five volumes, the fundamental nature of the physical processes that act on or just above the surface of the earth. The volumes in the series are *Climatology, Geomorphology, Biogeography, Hydrology* and *Soils*. The topics are treated in sufficient breadth and depth to provide the coverage expected in a *Fundamentals* series. Each volume leads into the topic by outlining the approach adopted. This is important because there may be several ways of approaching individual topics. Although each volume is complete in itself, there are many explicit and implicit references to the topics covered in the other volumes. Thus, the five volumes together provide a comprehensive insight into the totality that is Physical Geography.

The flexibility provided by separate volumes has been designed to meet the demand created by the variety of courses currently operating in higher education institutions. The advent of modular courses has meant that physical geography is now rarely taught, in its entirety, in an 'all-embracing' course but is generally split into its main components. This is also the case with many Advanced Level syllabuses. Thus students and teachers are being frustrated increasingly by lack of suitable books and are having to recommend texts of which only a small part might be relevant to their needs. Such texts also tend to lack the detail required. It is the aim of this series to provide individual volumes of sufficient breadth and depth to fulfil new demands. The volumes should also be of use to sixth form teachers where modular syllabuses are also becoming common.

Each volume has been written by higher education teachers with a wealth of experience in all aspects of the topics they cover and a proven ability in presenting information in a lively and interesting way. Each volume provides a comprehensive coverage of the subject matter using clear text divided into easily accessible sections and subsections. Tables, figures and photographs are used where appropriate as well as boxed case studies and summary notes. References to important previous studies and results are included but are used sparingly to avoid overloading the text. Suggestions for further reading are also provided. The main target readership is introductory level undergraduate students of physical geography or environmental science, but there will be much of interest to students from other disciplines and it is also hoped that sixth form teachers will be able to use the information that is provided in each volume.

John Gerrard

PLATES

All plates appear in a plate section between pages 112 and 113.

1   A red latosol, Tanzania
2   Yellow podzol with a distinct plough layer and organic staining, Tanzania
3   A podzol from the New Forest, England, showing a prominent bleached eluvial layer and iron pan
4   An earthy sulphuric peat soil with mineral layers
5   Extremely thin soil developed on rubbly chalk
6   Typical brown earth; fine loamy material over lithoskeletal siltstones and sandstones, Wyre Forest, England
7   Rendzina developed on limestone, Corfu
8   Blanket peat, Scotland, showing differing degrees of humification
9   Blocky structure with pronounced cracking in a red plateau soil, Tanzania
10 Termite mound, Tanzania
11 Ice wedge cast, Thetford Forest, eastern England
12 Well-developed stone line in red plateau soil, Tanzania
13 Frost hummock, Iceland, showing cryoturbated volcanic ash and windblown material
14 Wind eroded exposure, southern Iceland, showing prominent white and black volcanic ash layers and aeolian deposits
15 Well developed granular structure in topsoil
16 Blocky structure in topsoil

FIGURES

1.1   Basic subdivision of a soil profile   2
1.2   Demonstration of the way in which the pedosphere incorporates many elements from
      the other main geospheres   5
1.3   The hydrological cycle   8
1.4   The drainage basin hydrological cycle portrayed as a systems diagram   9
1.5   The carbon cycle   10
1.6   The nitrogen cycle   11
1.7   Composition by volume of a typical topsoil   12
1.8   Mineral particles, water films and active sites of respiration   17
1.9   The distribution of aerobic and anaerobic zones in soils under different drainage regimes
      (a) Well drained at field capacity and fully aerobic (b) Imperfectly drained at field capacity
      and with anaerobic microsites (c) Poorly drained at field capacity (d) Waterlogged soil   18
1.10 Distribution of water in soil   19
1.11 Arrangements of pedons and polypedons   19
2.1   Three major schemes of particle size distribution   23
2.2   Depiction of soil texture classes (a) British scheme (b) USDA scheme   24
2.3   Simplified triangular diagram of soil texture   24
2.4   The basic building block of silicate minerals, the silicon–oxygen tetrahedron   26
2.5   Tetrahedral linkage of silicate minerals (a) Single chain (b) Double chain   27
2.6   Representation of clay mineral structure (a) Biotite (b) Chlorite (c) Kaolinite
      (d) Vermiculite (e) Montmorillonite   28
2.7   Diagrammatic representation of main types of soil structure   32
2.8   Classification of soil fabric   35
2.9   Soil–water relationships   38
2.10 Clay–humus micelles and cation adsorption   42
3.1   Hypothetical evolution of a soil using the concepts of progressive and regressive
      pedogenesis   46

3.2   Development of eroded, composite and compound soil profile forms, Southern Alps,
      New Zealand   46
3.3   Illustration of soil formation as a combination of additions, losses, transfers and
      transformations   48
3.4   Loss-on-ignition values on samples from a peat bog at Ketilstaðir, south Iceland   49
3.5   Solubility of some common soil minerals as related to pH   52
3.6   Solubility of silica and alumina as related to pH   53
3.7   Organic constituents of a range of plant materials   57
3.8   Soil nitrogen transformations within soil   60
3.9   Sulphidisation of tidal marsh soils   61
3.10 Illustration of pedoturbation created by frost heave and the creation of frost circles 63
3.11 Involutions in the active layer   64
3.12 Typical infiltration curves for ponded infiltration on different materials   65
3.13 An example of discharge hydrographs for flow within different soil layers   65
3.14 Main soil transfer processes (a) Lessivage (b) Decalcification (c) Podzolisation
      (d) Gleisation (groundwater gley) (e) Gleisation (surface water gley)   67
4.1   The pedological hierarchy of soil factors, processes and materials   75
4.2   Idealised soil water balance   75
4.3   Annual soil water balance for (a) A hypothetical mid-latitude soil (b) A soil in
      northern England   76
4.4   Sediment yield as related to precipitation and latitude   77
4.5   Pedogenic accumulation rates as a function of effective moisture and atmospheric influx 77
4.6   Latitudinal variation in weathering types and depths   78
4.7   Relationship of soil orders to actual evapotranspiration and leaching index   79
4.8   (a) Clay accumulation (b) Iron accumulation with time in various environments   81
4.9   Weathering profile on granite rocks   87
4.10 Various ways in which water can move downslope   90
4.11 Soil gains and losses from different landform elements from near Saskatoon, Canada   92
4.12 Differentiation of mass movements on the basis of height and angle of failure   92
4.13 Generalised relationships between soil properties and slope angle   93
4.14 (a) Theoretical distribution of soil properties with slope form (b) The distribution
      of certain soil properties on chalk slopes   94
4.15 Rate of change of certain soil properties with time   94
4.16 Development of the weighted colour index values with time for a sequence of Holocene
      soils in the Wind River Mountains, Wyoming   96
4.17 Selected soil property chronofunctions illustrated with best-fit equations, Austerdalsbreen,
      Norway   97
4.18 Development of some soil orders with time   99
5.1   Possible horizon changes in a downslope direction   102
5.2   Master horizons   103
5.3   Development of surface horizons   104
5.4   Illustration of the subdivision of soil orders using spodosols as an example   110
5.5   Idealised soil profiles of the main Canadian soil orders   119
5.6   Distribution of nine soils from the Wyre Forest, England, on two principal component
      axes   125

5.7   Dendrogram for a single-linkage hierarchical classification of the nine soils portrayed in
      Figure 5.6   125
5.8   Dendrogram for a hierarchical classification produced by a centroid method   126
5.9   Hierarchical numerical classification of forty west Oxfordshire soils   127
5.10 Non-hierarchical grouping of soils from the Wyre Forest   128
5.11 Hierarchical grouping of soils from the Wyre Forest   128
6.1   Characteristic profile of a cambisol   132
6.2   Characteristic profile of a chernozem   134
6.3   Characteristic profile of a ferralsol   135
6.4   Characteristic profile of a luvisol   140
6.5   Characteristic profile of a phaeozem   141
6.6   Characteristic profile of a podzol   143
6.7   Characteristic profile of a placic podzol   144
6.8   Characteristic profile of a solonchak   147
6.9   Characteristic profile of a solonetz   149
7.1   The global distribution of Soil Taxonomy soil orders   153
7.2   Distribution of Soil Taxonomy soil orders on a hypothetical supercontinent   154
7.3   Precipitation map of Africa   157
7.4   Vegetation map of Africa   157
7.5   Generalised soil map of Africa   157
7.6   Generalised soil map of Australia   158
7.7   Generalised diagrammatic soil–landscape relations of the ridge-and-valley and
      adjacent plateau areas in central Pennsylvania   159
7.8   Section across the Stour valley, southeast England, showing the relationships of soil
      series to landforms and Quaternary deposits   160
7.9   Soil–site relationships on Dartmoor   162
7.10 Major soil types on the Dartmoor granite   162
7.11 Schematic representation of the synthetic alpine model   163
7.12 Some relationships of mean valley side slope to stream order   164
7.13 Variations in subaerial processes with slope component and stream order, Queensland,
      Australia   164
7.14 Dirichlet tesselation of the soils of the Wyre Forest (a) The six groups of the optimal
      classification (b) The groups created by a spatial weighting of 10 (c) The groups created
      by a spatial weighting of 17.5   166
7.15 Maps of percentage sand and percentage clay in the subsoil by kriging on a
      20 × 25 m grid. Bottom: map of soil types on the two sides of a long transect   167
7.16 Hypothetical nine-unit land-surface model   168
7.17 Soil and vegetation catena representative of the tundra near Prudhoe Bay, northern Alaska   172
7.18 Landforms and deposits of a typical floodplain   173
7.19 Glacial depositional landscape sequence   174
8.1   Main interacting factors involved in soil degradation   178
8.2   Universal soil loss equation nomograph for calculating soil erodibility   186
8.3   Movement of nitrate as a pulse   192
8.4   Soil compaction and the formation of cultivation pans   198

TABLES

1.1	Geological classification of the most commonly occurring bedrock parent material	6
1.2	An alternative rock classification based on texture	7
2.1	Key for the assessment of soil texture	25
2.2	Some properties of the more commonly occurring clay minerals	29
2.3	Some typical bulk densities and porosities of soils	30
2.4	Soil temperature regimes	37
3.1	Classification of living organisms on the basis of energy and carbon nutrition	56
4.1	Zonal soils as classified by Dokuchaev	73
4.2	Clay minerals generally associated with different climatic zones	80
4.3	Input rates of litter under differing vegetation cover	83
4.4	Weathering sequence for common rock-forming minerals	85
4.5	Relations between landscape components and some soil moisture characteristics	91
4.6	Dominant factor explaining the variation in soil properties on chalk slopes	93
5.1	Soil properties on stepped erosion surfaces in Iowa	105
5.2	Classification of soils by Dokuchaev	107
5.3	Soil Survey Staff soil orders, formative elements and connotations	110
5.4	Major FAO/UNESCO soil groupings, formative elements and connotations	114
5.5	Approximate relationships between the Soil Survey Staff soil orders and FAO/UNESCO major soil groupings	115
5.6	Classification of soils according to the scheme adopted in the British Isles and US Soil Taxonomy and FAO/UNESCO equivalents	117
5.7	Soil order nomenclature in the Australian soil classification scheme	121
7.1	Relationships of Soil Survey Staff soil orders to broad bioclimatic zones	155
7.2	Relationships between FAO/UNESCO soil orders and climatic regimes	156
7.3	Plate segments and characteristics	156
7.4	Summary characteristics of the six groups of the optimal classification of soils of the Wyre Forest	165

7.5   Relationships between soils, stream order and the units of the nine-unit land-surface
       model   170
7.6   Classification of tropical savanna catenas   171
8.1   Proportions of stable, soil-degraded and non-used land by continent   180
8.2   Soil degradation degree by continent   181
8.3   Most commonly used soil erodibility indices   187
8.4   Relative erodibility of 4–5mm aggregates from different slope positions under forest and
       farmland   188
8.5   Composition of aggregates under forest and farmland   189
8.6   Recommended upper limits of total metals in soils   197
8.7   Dutch guidelines for several soil metal pollutants   197

BOXES

1.1	Characteristics of systems	3
1.2	Two-dimensional soil–slope system	20
2.1	A scheme for describing soil consistency	36
3.1	The more commonly used chemical weathering indices	55
3.2	Van Post scale of peat decomposition	58
3.3	Organic theory of podzol formation	68
3.4	Inorganic theory of podzol formation	69
4.1	State factor equations	73
4.2	Granite weathering profiles	87
4.3	Weathering profile for chalk	88
4.4	Weathering profile on shales	88
4.5	Relationship between texture and landforms in the Indus valley	89
4.6	Some empirical relationships between soil transport and slope length and slope angle	91
4.7	Soil age and characteristics in New Mexico	99
5.1	Horizon boundaries	102
5.2	Master horizon qualifications	106
5.3	Surface diagnostic horizons (epipedons)	108
5.4	Subsurface diagnostic horizons	109
5.5	Representation of texture pattern in the scheme devised by FitzPatrick	123
7.1	Development of soils in hummocky disintegration moraines	175
8.1	GLASOD soil degradation types	178
8.2	Classification of severity of soil degradation	181
8.3	Soil erosion in India	182
8.4	Main factors affecting soil erosivity and erodibility	183
8.5	GLASOD methodology for assessing degree of water erosion	184
8.6	Types of soils susceptible to, or resistant to, specific forms of water erosion	185
8.7	Various attempts at quantifying rainfall erosivity	187
8.8	GLASOD methodology for assessing wind erosion	190
8.9	GLASOD methodology for assessing degree of nutrient depletion for soils in dryland areas	191

ACKNOWLEDGEMENTS

Thanks are due to the following authors, publishers and learned bodies for permission to reproduce figures and tables:

Fig. 7.19 from D.E. Sugden and B.J. John (1976), *Glaciers and Landscape*; Table 8.1 and Boxes 8.2, 8.8, 8.9 from N. Middleton and D. Thomas (1997), *World Atlas of Desertification*, by permission of Edward Arnold Ltd; Table 5.6, CAB International; Fig. 3.2 from P.J. Tonkin and L.R. Basher (1990), 'Soil-stratigraphic techniques in the study of soil and landform evolution across the Southern Alps, New Zealand', *Geomorphology* 3: 547–75, with permission from Elsevier Science; Fig. 4.13 from Furley, P.A. (1968), 'Soil formation and slope development: 2. The relationships between soil formation and gradient in the Oxford area', *Zeitschrift für Geomorphologie* NF 12: 25–42, by permission of the editor and Gebruder Borntraeger; Figs 5.6, 5.7, 5.8, 5.9, 5.10, 5.11 from R. Webster and M.A. Oliver (1990), *Statistical Methods in Soil and Land Resource Survey*, by permission of Oxford University Press; Tables 2.1, 7.2 and Fig. 5.4 from E.A. Fitzpatrick (1983), *Soils: Their Formation, Classification and Distribution*; Tables 2.2, 2.3 and Figs 1.8, 1.9 from D.L. Rowell (1994), *Soil Science: Methods and Applications*, reprinted by permission of Pearson Educational Ltd; Table 4.6 and Fig. 4.14 from P.A. Furley (1971), 'Relationships between slope form and soil properties developed over chalk parent materials', in *Slopes, Form and Process*, ed. D. Brunsden, IBG Spec. Pub. 3, by permission of the Royal Geographical Society (with Institute of British Geographers); Tables 8.4, 8.5 from A.C. Imeson and P.D. Jungerius (1976), 'Aggregate stability and colluviation in the Luxembourg Ardennes', *Earth Surface Processes and Landforms* 1: 259–71, by permission of J. Wiley and Sons; Fig. 7.2 from A.H. Strahler and A.N. Strahler (1992), *Modern Physical Geography*, 4th edn; and Figs 7.3, 7.4, 7.5, 7.6 from H.D. Foth and J.W. Schafer (1980), *Soil Geography and Land Use*, by permission of J. Wiley and Sons, New York.

The author wishes also to thank Margaret Oliver for her work in preparing the colour plates.

Every effort has been made to contact copyright holders for their permission to reprint material in this book. The publishers would be grateful to hear from any copyright holder who is not here acknowledged and will undertake to rectify any errors or omissions in future additions of this book.

1

THE NATURE OF SOIL

WHAT IS SOIL?

There are many different opinions as to what constitutes soil, and there is no commonly agreed definition. Soil is the superficial covering of most of the land area of the Earth and varies in thickness from a few millimetres to many metres. One definition is that soil is weathered material at the Earth's surface which may or may not contain organic matter and often also contains air and water. A more all-embracing definition is that soil is a natural body composed of minerals, organic compounds, living organisms, air and water in interactive combinations produced by physical, chemical and biological processes. This would be a genetic soil definition. The primary components of soil are inorganic materials, mostly produced by the weathering of bedrock or other parent material, various forms of organic matter, gas and water required by plants and soil organisms, and soluble nutrients used by plants. These constituents differ from the parent material in their morphology, physical, chemical and mineralogical properties, and their biological characteristics.

There is often confusion with the term **regolith**. According to Bates and Jackson (1987), 'regolith' is the layer or mantle of fragmented and unconsoli-

dated rock material, whether residual or transported and of highly varied character, that nearly everywhere forms the surface of the land and overlies or covers the bedrock. It includes rock debris of all kinds, volcanic ash, glacial drift, alluvium, loess and other aeolian deposits, vegetal accumulations, and soil. Ollier and Pain (1996) use 'regolith' as a composite term for a variety of earth materials that may be closely related. These materials are *weathered rock*, which, if it is in place, is called saprolite, *residuum*, which is weathered material that has been disturbed and moved, *transported surficial sediments*, *chemical products* in the near-surface environment, *soils*, and *miscellaneous products*, including volcanic ash and lag gravels.

Soil has a definite three-dimensional organisation. In a vertical direction, soil constituents may be arranged into horizons of mineral and/or organic constituents of variable thickness. A simple sub-division into four horizons is often made (Figure 1.1). The processes creating such horizons are examined in Chapter 3 and their detailed characteristics in Chapter 5. Soil grades at its lower margin to hard rock or materials usually devoid of the marks of biological activity. Thus, the lower limit is normally the limit of biological activity and, according to Soil Survey Staff (1975, 1994), this

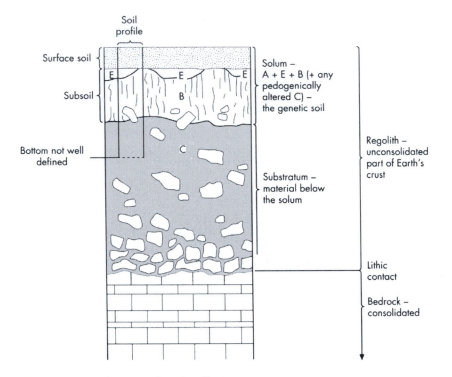

Figure 1.1 Basic subdivision of a soil profile.

generally coincides with the common rooting depth of native perennial plants. However, the bottom of the genetic soil may not coincide with this rooting depth, being either shallower or deeper depending on circumstances. In the recent Australian classification, the term 'pedologic organisation' is used (McDonald *et al.*, 1990). This concept includes all changes in soil material resulting from the effect of the physical, chemical and biological processes that are involved in soil formation. Whatever approach is adopted, the lower limit of a soil is always difficult to establish, and this confusion draws attention to the somewhat arbitrary distinction between soil and regolith. A solution to the dilemma may be to use the term 'solum' for the genetic soil developed by soil-building processes. This is the manner in which the term 'soil' will be used in this book.

In lateral directions, soil varies in response to variations in the major soil-forming factors such as climate, relief and parent material. Soil forms part of the landscape, and the term 'soilscape' is sometimes used. This is implicit in the concept of soil geomorphology (Gerrard, 1992a), which is an assessment of the genetic relationships between soils and landforms. Soil responds to and influences environmental processes and conditions. It is also very vulnerable. The top part of the soil profile is constantly being removed, added to or generally reworked. Also, there are continual additions and removals within the soil as a result of vertical and lateral water movement. Soil is a vital natural resource and is the medium within which most agriculture takes place. Thousands of years of agriculture and settlement mean that there are very few 'natural' soils left on the Earth's surface, and there are many areas of the world where soils are being degraded by human activities. Thus it is very important to understand the basic characteristics of soils and the way in which those characteristics are shaped by specific processes.

SOILS AS SYSTEMS

Soils are the result of the interaction of many processes both within and outside the soil body, and it is necessary to treat soils as open systems, losing and receiving material and energy at their boundaries. Thus:

> soils are complex open process and response systems. As such they continuously adjust by varying degrees, scales, and rates to constantly changing energy and mass fluxes, thermodynamic gradients and other environmental conditions, to thickness changes, and to internally evolved accessions and threshold conditions. Consistent with these facts is the notion that disturbance and change is a natural, predictable consequence of all soil evolving processes.
>
> (Johnson and Watson-Stegner, 1987: p. 363)

A great many models of soil formation have been suggested, but they tend to fall into one of three general approaches. These are the functional factorial approach (Dokuchaev, 1898; Jenny, 1941), the system–process flux approach (Simonson, 1978), or some synthesis of both approaches (see Chapter 4). Although models of soil formation vary, they are all based on the notion that soils function as open systems. The essential characteristics of systems are listed in Box 1.1.

Systems may be classified in several ways. Morphological systems are the instantaneous properties that, in their organisation, form the observable part of reality. Soil as a morphological system would be what is called the soil **skeleton**. The skeleton is composed of the relatively stable and not easily translocated mineral grains and resistant organic bodies. Cascade systems are composed of a chain of subsystems that are linked dynamically by flows of mass or energy. In soil, this is represented by the soil **plasma**, that part of the soil capable of being moved, reorganised and concentrated. It is the active part of the soil and includes all material, mineral or organic, of colloidal size and the relatively soluble material. Integration of the morphological and

---

Box 1.1

CHARACTERISTICS OF SYSTEMS

1 Systems possess boundaries, either real or arbitrary.
2 Systems possess inputs and outputs of energy and matter crossing these system boundaries.
3 Systems possess pathways of energy transport and transformation associated with matter within the system.
4 Within systems, matter may be transported from place to place or have its physical properties transformed by chemical reaction or change of state.
5 Open systems tend to attain a dynamic equilibrium or steady state in which rate of input of energy and matter equals rate of output of energy and matter, while storage of energy and matter remains constant.
6 Where input or output rates of an open system change, the system tends to achieve a new dynamic equilibrium. The period of change leading to the establishment of the new equilibrium state is a transient state and the period of time involved will depend on the sensitivity of the system.
7 The amount of storage of energy and matter increases (decreases) when the rate of energy and material flow through the system increases (decreases).
8 The greater the storage capacity within the system for a given input, the less is the sensitivity of the system.

cascade systems produces process–response systems, which represent the totality of the soil system. It is worth mentioning a fourth type of system, which is a control system, where there is some external control such as when soil is managed for cultivation.

The operation of soil systems can be examined in varying amounts of detail. At the simplest black box level the entire system is examined as a unit with no consideration of internal structure or what happens within the system. An example of this approach would be the measurement of precipitation as input into the soil and water emerging as output at the soil base without considering pathways, stores or lags. At a grey box level, a partial view of the system is examined. At this level, the soil body might be recognised as a potential regulator of water movement and perhaps also as having a storage capacity. The most complex and most realistic treatment of the soil body is as a white box, where as many of the regulators, stores and flows as possible are identified and analysed. Extending the water movement analogy, movement in individual soil layers would be considered at the white box level of analysis. The soil column can be considered to be a series of compartments through which there is continuous movement of material. Material is removed from the system by vegetation uptake, erosion and drainage, and is added to the system by atmospheric inputs, and from animal and vegetation activity. It is also recognised that additions and removals can occur from, and to, adjacent soil columns. The ways in which particular soils are formed can only be considered by a white box approach.

The open system nature of soils and the way in which systems react to change (see items 5 and 6 in Box 1.1) can be seen with respect to the thickness of soil at any one point. This thickness will depend on the relative rates of soil formation and soil removal. At some places, soil removal (erosion) is slight and deep soils develop; at other sites, greater rates of soil removal leads to thin soils. This can be thought of in terms of a balance, sometimes called a **denudational balance** but more correctly called a 'soil-denudational balance'. Soils can increase in depth by *in situ* production of material by weather-

ing and soil formation and by inflow of material, usually on the surface from upslope. In many parts of the world, aeolian input is a significant factor. Deposition of fine volcanic ash can also be important locally. Removal of material is usually thought of as surface removal by slopewash, deflation and mass movements, but material can be removed from within soils by water movement and a variety of animals and organisms. The balance between these various inputs and outputs can lead to three alternative situations:

$$A = R; \quad A < R; \quad A > R$$

where A represents all additions to the soil profile and R represents removals. Soil thickness will remain constant or increase or decrease according to the relative strengths of the respective processes. If removal processes are more rapid than weathering, soil formation and surface additions, then only a thin soil will exist. Soils will be kept thin and permanently youthful or undeveloped. If rates of weathering and soil formation are more rapid than removal processes then a thick soil will develop. Weathering and soil formation tend to be at a maximum at intermediate soil thicknesses. When soils are thin, little water is retained and soil formation is slow, whereas in thick soils water may move so slowly towards bedrock that the rate of weathering and the production of new soil material is inhibited.

In considering soil as an open system, two types of question immediately need answering. The first question seeks answers to such problems as how the system originated and developed. An answer to such a question, with respect to soil, must include the sequence of events that affected the soil and the processes involved in its development. Soils possess a history, which must be considered in understanding their characteristics. They might have experienced major climatic or vegetation changes and been subject to extensive periods of erosion or addition of surface material. This also means that many of the characteristics of soils may possibly be used to infer past conditions, but this will only be achieved if we know the answer to the second type of question, namely how the system works or functions. This asks for an understanding of the status and role of the many forces and processes acting on and within the soil system.

GEOSPHERES

It is clear from the preceding discussion that soils do not exist in isolation. The layer of soil on the Earth's surface is often called the **pedosphere** and is only one of several **geospheres** that can be defined. The term 'geosphere' is usually used in the sense advocated by Friedman (1985) to include the totality of the geophysical systems. However, Ellis and Mellor (1995) restrict the term 'geosphere' to the solid, upper part of the crust in which soils are formed. 'Geosphere' is used here in the manner suggested by Friedman. A simple analysis of geospheres indicates that the pedosphere interacts with the other main geospheres, namely the **atmosphere**, **biosphere**, **lithosphere** and **hydrosphere** (Figure 1.2). Thus the pedosphere is the zone of interaction of all these geospheres, and it is not surprising that

when the main factors controlling soil formation are examined (Chapter 4), such factors belong to one or other of these geospheres. For a more thorough analysis of geospheres, the reader is directed to Huggett's (1995) book.

Atmosphere

The atmosphere is the layer of air surrounding the Earth. It is a dusty gas, much of the mass of which is contained in its lowest layer. The air in the atmosphere is dominated by two gases – nitrogen (about 78 per cent by weight) and oxygen (about 21 per cent). The air also contains water vapour, other gases such as carbon dioxide and argon, and suspended particulate matter. It is usually divided into several 'spheres', each with its own characteristic temperature, pressure and composition. These spheres are the troposphere, stratosphere, mesosphere, thermosphere and exosphere. The ionosphere, a shell of high electron concentration extending out into space, could also be included. Weather and climate, which interact with the pedosphere, are confined to the relatively dense lower level, or troposphere. The troposphere extends to a height of approximately 16 km at the equator and 8 km at the poles. Temperatures in the troposphere decrease at a rate of about $6.5°C\ km^{-1}$. The stratosphere extends to an altitude of about 50 km. Temperatures increase through this layer as a result of the absorption of ultraviolet radiation from the sun by ozone.

Biosphere

The biosphere encompasses the fauna and flora that live above, at and below the Earth's surface. It includes a vast variety of organic compounds and life forms, comprised essentially of hydrogen, oxygen and carbon, as well as dead organic matter. The zone of interaction between the biosphere, lithosphere, atmosphere and pedosphere is sometimes called the ecosphere. The ecosphere can be divided into fundamental systems called **ecosystems**. Perhaps the most all-embracing definition of biosphere is 'the integrated living and life-supporting system

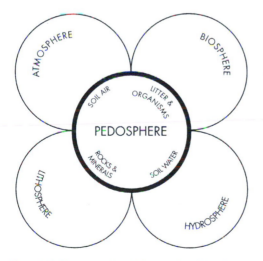

Figure 1.2 Demonstration of the way in which the pedosphere incorporates many elements from the other main geospheres.

comprising the peripheral envelope of Planet Earth together with its surrounding atmosphere so far down and up, as any form of life exists naturally' (Polunin and Grinevald, 1988: p. 118).

Lithosphere

The lithosphere is the uppermost layer of the Earth's crust and is the initial parent material on which soils develop. The rocks making up the lithosphere can be subdivided into igneous, metamorphic and sedimentary rocks. Igneous rocks are the primary rocks, and their eroded and weathered products are transported and then deposited, eventually to form sedimentary rocks. Metamorphic rocks are altered igneous and sedimentary rocks. The specific characteristics of individual rock types determine their weathering susceptibility and the physical and chemical nature of the soil mineral component.

Igneous rocks, formed by the cooling and solidification of molten magma, can be differentiated according to their mode of emplacement. Plutonic rocks, such as granite, have cooled slowly at great depths, allowing the rock to become coarse-grained. Incorporation of siliceous material from the Earth's crust also ensures that they are acidic in composition. Hypabyssal rocks, such as dolerite, have cooled at intermediate depths and are medium-grained in texture. Volcanic rocks are the extrusive igneous rocks formed on the Earth's surface. They are essentially fine-grained and less acidic (more basic), because the magma has come straight from deep within the Earth and contains less silica and more ferromagnesian minerals. Basalt is a good example. A variety of ejected pyroclastic rocks, which solidify as volcanic breccias, can also be formed.

Sedimentary rocks vary considerably in grain size, texture, cohesion and chemical composition as a result of the nature of the rock from which they are derived and the environment of deposition. In terms of grain size, a distinction is made between rudaceous, arenaceous and argillaceous rocks. Rudaceous rocks are composed of coarse angular (breccias) or rounded (conglomerates) rock fragments, bound together by calcareous, siliceous or some other matrix. Arenaceous rocks are sandstones and argillaceous rocks are clays, shales, marls and mudstones with a large proportion of clay-sized particles. Other sedimentary rocks, such as limestones and coal, are classified on the basis of chemical composition.

Metamorphic rocks have been formed from igneous and sedimentary rocks by intense processes involving heat, pressure and chemical alteration. The most abundant are gneiss, schist, slate and quartzite. The main rock types are listed in Table 1.1.

Sedimentary rocks account for about 75 per cent of the exposed land surface of the world, intrusive igneous rocks comprise 15 per cent, extrusive rocks 3 per cent and metamorphic rocks the rest. An alternative classification, based on texture, might be more appropriate when considering rock as a parent material for soil formation (Table 1.2). Rock texture will determine some of the characteristics of the weathering products. Mineral types and chemical elements are also important. Quartz and feldspars are the dominant minerals and probably exist in approximately equal amounts (feldspar 30 per cent, quartz 28 per cent); calcite and dolomite make up 9 per cent, and clay minerals and micas about 18 per cent of surface material.

Unconsolidated superficial deposits are the parent material of many soils and can be classified according to the nature of their depositional environment, for example riverine alluvium, glacial, marine,

Table 1.1 Geological classification of the most commonly occurring bedrock parent material

Igneous	Sedimentary	Metamorphic
Andesite	Chalk	Gneiss
Basalt	Dolomite	Hornfels
Diorite	Gritstone	Marble
Dolerite	Limestone	Phyllite
Granite	Mudstone	Quartzite
Gabbro	Sandstone	Schist
Rhyolite	Shale	Slate
Syenite	Siltstone	

Table 1.2 An alternative rock classification based on texture

Texture	Rock type
Crystalline	
Soluble carbonates	Limestones, dolomite, marble
Banded silicate minerals	Gneiss
Randomly oriented and distributed silicate minerals, reasonably coarse-grained	Granite, diorite, gabbro, syenite
Randomly oriented and distributed silicate minerals in background of fine grain	Basalt, rhyolite and other volcanic rocks
Clastic	
Stably cemented	Silica- and iron-cemented sandstone
Slightly soluble cement	Calcite-cemented sandstone
Highly soluble cement	Gypsum-cemented sandstone
Weakly or incompletely cemented	Friable sandstones, tuff
Uncemented	Clay-bound sandstone
Very fine grained	
Isotropic, hard rocks	Hornfels, some basalts
Anisotropic on a macro scale but microscopically isotropic hard rocks	Cemented shales
Microscopically anisotropic hard rocks	Slate, phyllite
Soft, soil-like rocks	Compaction shale, chalk, marl
Organic rocks	Lignite, bituminous coal

Source: After Goodman (1980)

lacustrine, aeolian or slope deposits (colluvium). These deposits may be cemented or indurated by calcareous, silicic or iron-rich compounds.

Hydrosphere

The hydrosphere includes all the forms in which water can occur on or below the Earth's surface. It includes water passing through or held in the soil, and water in rivers, lakes, underground storage and oceans. It also includes the chemical composition of water, essentially hydrogen and oxygen, but with many other elements in varying quantities and combinations. Sea water contains predominantly sodium and chloride ions, approximately 85 per cent by weight, producing a mean salinity of $35 \, g \, kg^{-1}$. The chemical composition of terrestrial water varies quite considerably depending on the chemicals it encounters.

It is estimated that approximately $1.4 \times 10^9 \, km^3$ of water exists in various states, most of it stored in the oceans. Only 2.6 per cent of the hydrosphere is fresh water, and of this, 77.2 per cent is frozen in ice caps, icebergs and glaciers. Groundwater accounts for about 22.2 per cent of fresh water.

Pedosphere

The pedosphere must be the layer on the surface of the Earth in which soil-forming processes operate. In this book, it is regarded as the solum, defined earlier. Huggett (1995) has suggested the term 'edaphosphere' for this layer and the general term 'debrisphere' for all material lying on the lithosphere. But 'pedosphere' is a perfectly acceptable term, used by most pedologists, and is the context in which it is used here.

Toposphere

An additional sphere, the toposphere, is sometimes included in geospheres. The toposphere 'sits at the interfaces of the pedosphere and atmosphere and pedosphere and hydrosphere' (Huggett, 1995: p. 7). It is really a surface of interaction in only two dimensions, unlike the other spheres, which possess three dimensions. However, there is no doubt that it is an important zone of interaction, especially between geomorphological and pedological processes. It is essentially the 'pedomorphic surface' (Dan and Yaalon, 1968) or the 'pedoderm' (Brewer *et al.*, 1970) and the main focus of soil geomorphology.

GLOBAL CYCLES

The interaction between the pedosphere and the other spheres can be seen in a series of global cycles.

The ones examined here are the hydrological cycle, carbon cycle and nitrogen cycle.

Hydrological cycle

The hydrological cycle includes the continuous interchanges between the hydrosphere and all the other spheres (Figure 1.3). The components of the hydrological cycle in relation to the soil (pedosphere), expressed in systems terminology, can be examined in greater detail (Figure 1.4). What is of interest here is the relative volume and the pathways that water takes. The nature and extent of vegetation cover will determine how much precipitation actually reaches the soil. Much will be intercepted and lost through evapotranspiration. Some will reach the soil surface as direct throughfall, but some will be transferred from the vegetation as stemflow and drip flow. In this process, the chemical composition of the water may be altered.

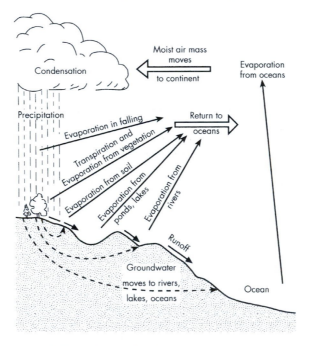

Figure 1.3 The hydrological cycle.

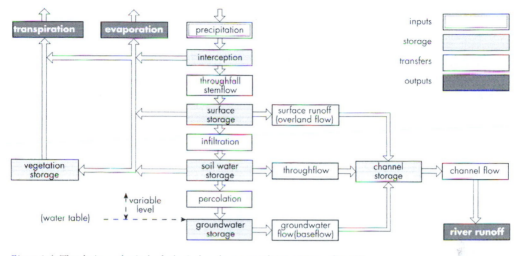

Figure 1.4 The drainage basin hydrological cycle portrayed as a systems diagram.

Precipitation reaching the surface will accumulate as depression storage and then either run off as surface flow or infiltrate into the soil. The relative amounts following each path will depend on surface characteristics and soil properties. Once in the soil, water will either percolate vertically towards the water table or move laterally as throughflow. Water in the groundwater zone will move slowly towards the nearest river, stream, lake or sea.

Carbon cycle

The **carbon cycle** refers to the movement of carbon throughout the various 'spheres' (Figure 1.5). Most carbon is stored as carbonate sediments in the lithosphere. Only about 0.2 per cent is readily available to organisms as carbon dioxide or as decaying biomass. In the atmosphere, carbon moves mainly as carbon dioxide. It is also dissolved in water on land and in the oceans. Carbon also resides in carbohydrate molecules in organic matter, as hydrocarbon compounds in rock (petroleum, coal) and as mineral carbonate compounds such as calcium carbonate. Carbon dioxide in the atmosphere is provided by respiration from plant and animal tissues on land

and in the oceans. Some new carbon can enter the atmosphere from volcanoes in the form of carbon dioxide and carbon monoxide. Recently, there has been much concern about the production of carbon dioxide and carbon monoxide through the combustion of fossil fuels and their role, known as the enhanced greenhouse effect, in global warming. Carbon dioxide leaves the atmosphere and enters the oceans, where it is used by phytoplankton. These organisms are consumed by marine animals that also build skeletal structures of calcium carbonate. This mineral matter settles on the ocean floors to accumulate as sedimentary strata, thus continuing the cycle. Soils are important 'sinks' for carbon dioxide. The movement of carbon through, and its transformation within, the soil is extremely important.

Nitrogen cycle

The nitrogen cycle is also important for what happens in the soil (Figure 1.6). Like carbon, nitrogen moves through the global cycle in a variety of forms. The main source is nitrogen in the atmosphere, which has to be 'fixed' before it can become

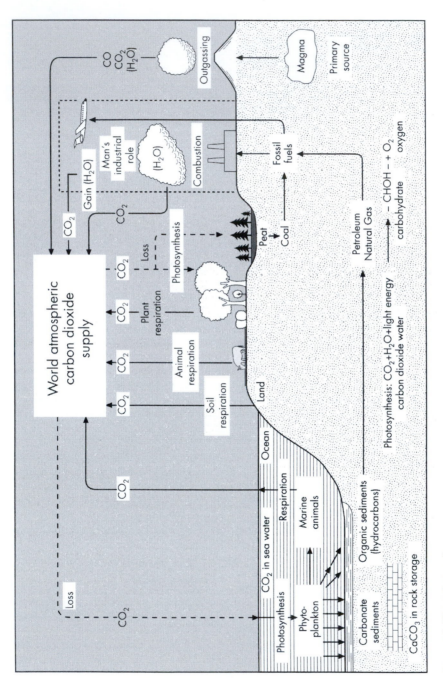

Figure 1.5 The carbon cycle.

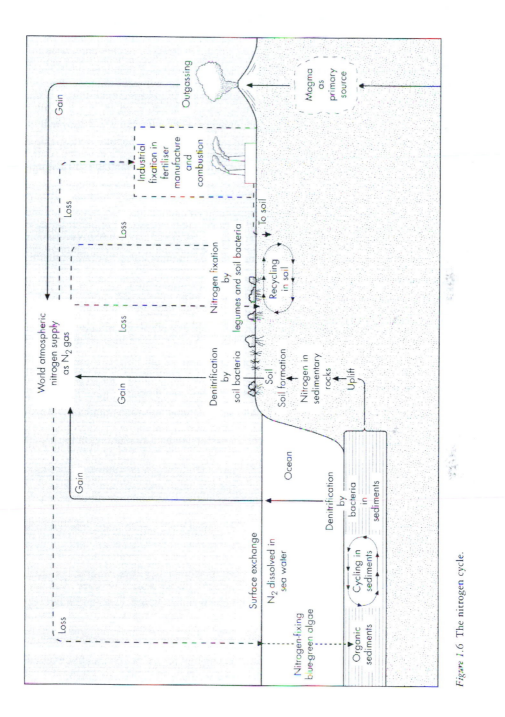

Figure 1.6 The nitrogen cycle.

assimilated by plants or animals. Some species of soil bacteria can 'fix' nitrogen, and some cyanobacteria can also fix nitrogen. There are also symbiotic relationships between nitrogen-fixing bacteria of the genus *Rhizobium* and some species of trees and plants. Almost all members of the legume family are involved in such relationships. *Rhizobium* bacteria infect the root cells of these plants in root nodules produced jointly by the plant and the bacteria. The bacteria supply nitrogen to the plant through fixation, while the plant supplies the nutrients and organic compounds needed by the bacteria. Nitrogen is lost from the soil and biosphere by the process of denitrification (see Chapter 3), whereby nitrogen in usable forms is converted back into nitrogen and other unusable forms.

THE SOIL SYSTEM

The previous sections have stressed that soils behave as systems with a good level of internal organisation. This organisation is reflected in the soil properties. The striking characteristics of soils, such as colour, texture, structure and thickness, are due to the combinations and arrangements of inorganic and organic soil constituents with specific physical and

chemical properties, and voids filled with either air or water. These characteristics allow one type of soil to be distinguished from another and form the basis of soil classification schemes. They also determine whether a soil is fertile or infertile, dry or water-logged, and the use to which that soil may be put. The inorganic material is composed of mineral particles derived either from parent material by weathering or from a depositional input at the surface. These mineral particles will also be altered by weathering within the soil. Organic soil material is derived from living and dead organisms. It is the general nature of soil constituents that is examined here. Specific soil properties are examined in the next chapter. The main constituents of soil are mineral particles, organic compounds and living organisms, air, and water (Figure 1.7).

Mineral particles

Mineral matter forms a large proportion of the soil body. The nature and kinds of mineral particles vary within soils and from place to place. As will be seen in Chapter 4, the parent material on which the soil has developed probably has the greatest influence on the nature of the mineral particles. This parent material will usually be the bedrock at the base of the soil profile, but many soils develop on trans-ported material, and it is this transported material that will impart characteristics to the soil. Some examples of this are soils developed on thick loess deposits, on river alluvium, on glacial deposits and on sand dunes. Also many soils on hillslopes in temperate regions have developed on a layer of soliflucted material formed under periglacial conditions associated with the last ice age. Such soliflucted material may or may not reflect the nature of the underlying bedrock. Mineral matter may also enter the soil from the surface. It is now recognised that most soils contain quantities of windblown material. Even the British Isles receives, from time to time, dust and fine sand blown from the Sahara Desert. Areas adjacent to deserts and regions where wind erosion is prevalent can contain

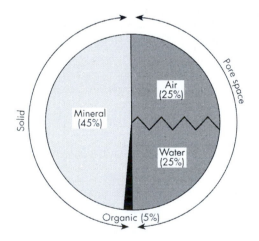

Figure 1.7 Composition by volume of a typical topsoil.

appreciable quantities of aeolian material. Volcanic ash or tephra is also an important constituent in some soils. In Iceland, soils have been affected substantially by the input of both windblown silt and fine sand, and of volcanic ash.

In many cases, the nature of soil mineral particles can be related to the nature of the underlying bedrock, and differences in bedrock types help to explain differences in soil types. A soil developed on limestone will be composed of different minerals and materials to one developed on granite (see Chapter 4). Soils developed on sandstones will differ yet again. Mineral particles will vary in size, shape and their chemical composition. Some will be single mineral particles, but others will still be incompletely weathered aggregates composed of more than one mineral. The importance of mineral particles is that they are 'building blocks', the components that impart structure and, to a certain extent, strength. They also influence the operation of many soil processes.

Mineral particles are produced within the soil by three main groups of processes: physical, chemical and biological. *Physical processes* result in the disintegration of parent material without any change in chemical composition. The mineral particles are simply broken-down versions of the original material. Such disintegration is the result of thermal expansion and contraction, wetting and drying, freeze–thaw activity and possibly salt weathering. Attrition of particles as a result of differential movement will also have an effect. *Chemical processes* produce altered or completely new minerals. Unchanged rock minerals are called inherited or primary minerals, and altered or new minerals are called pedogenic or secondary minerals. *Biological processes* act mainly by producing weak acids in solution, which aid the chemical decomposition of minerals. Although the processes acting around plant roots are predominantly chemical, the physical action of growing roots and burrowing organisms may also aid the physical breakdown of mineral particles.

Organic constituents

Organic compounds

The main organic constituent of soils is humus. Although it is distributed throughout the soil, humus is mostly concentrated in the topsoil. Humus is the semi-decayed remains of plant and animal life, which ultimately breaks down into organic compounds. It contains colloidal substances, which aid the transfer of nutrients to plants. Organic materials promote water retention and provide an environment for living organisms. Soils with high humus contents are usually chemically active, with high exchange capacities (see Chapter 2). Soils poor in organic matter are 'inactive', with slower chemical actions and reactions. Much will depend on the nature of the vegetation from which the humus was derived and the speed of the breakdown processses. Tropical rainforests possess a magnificent vegetation cover of great mass and variety, but soils developed under such vegetation are low in organic matter because of high rates of decay and the quick uptake of the organic residue by living plants. In contrast, the decay of organic residues from conifers is extremely slow and produces high acidity. This reflects partly the nature of the vegetation and partly the climatic environment within which most conifers grow. Organic compounds are important to the chemical and biological activity of soils, and the amount and distribution of organic matter often allows one soil to be distinguished from another.

Living organisms

It is easy to forget that soil is the habitat of millions of organisms and that these living organisms play a vital role in the functioning of the soil system. They break down and transport constituents from one part of the soil to another. They aerate the soil, increase its porosity and add organic acids and carbon dioxide to the soil. These organisms derive energy for growth from the oxidative decomposition of complex molecules. During decomposition, there is a conversion from organic forms to inorganic forms such as ammonium, phosphate and sulphate. This process is known as mineralisation. Some of this material is immobilised by incorporation into the cell substances of soil organisms. This is essentially the carbon cycle described previously.

Soil organisms can be classified as producers, consumers or decomposers. Producers, such as plants, fix carbon from atmospheric carbon dioxide during photosynthesis, consumers feed on plants and other organisms and decomposers utilise carbon from organic material. One cubic foot (0.026 m^3) of organic soil in a temperate woodland is thought to contain 45 million single-celled protozoa, 4 million subinsects (such as eelworms and rotifers), 60,000 other insects, including larvae, ants and beetles, also millipedes and woodlice, and 150 earthworms. These creatures will have a combined weight of 120 g. In addition, there may be up to 300 g of fungi, moulds and algae, up to 900 g of roots of higher plants, and 1.3 million bacteria present per ounce. The weight of organisms is known as **biomass**. Microbial biomasses are usually expressed as a weight of organic carbon per hectare ($kg \text{ ha}^{-1}$). In arable soils, the microbial biomass is approximately 2 per cent of soil organic matter. It is approximately 3 per cent in grassland and woodland soils.

A distinction can be made between organisms on the basis of size. Macrofauna ($> 1,000 \mu m$) include all the larger vertebrate animals, such as rabbits, moles and so on, that live or burrow in the soil. Mesofauna ($200–1,000 \mu m$) are organisms that are usually visible to the naked eye and include earthworms, ants, termites, beetles and so on. Animals, plants and fungi that are not visible to the naked eye are called micro-organisms ($< 200 \mu m$), or microfauna and microflora.

Some of the mesofauna, such as earthworms, pass organic and mineral substances through their bodies, producing faecal materials that alter soil texture and perhaps change a soil's chemical behaviour. Faecal material is not permanent but is usually broken down by micro-organisms. Material that has passed through the digestive tracts of organisms has had its physical and chemical prop-erties altered, sometimes in very subtle ways. The mixing of material by these larger organisms will tend to blur any soil horizons that might be developing. It is no coincidence that soils within which mesofauna are active possess the least distinct soil horizons.

Mesofauna can be divided into four main groups: arthropods, nematodes, annelids and molluscs. Arthropods – mites, springtails, beetles and many insect larvae, ants termites, millipedes and centi-pedes – usually feed on detritus in the lower part of the litter layers. Many species of beetle live in the soil. Perhaps the most useful are coprophagous beetles, which feed on the dung of large herbivores and increase the rate at which it is mixed with the soil. Termites comminute all forms of litter, trunks and branches of trees and leaves. Most are surface feeders and build termitaria. Centipedes are carnivorous, and millipedes feed on vegetation.

Nematodes are unsegmented worms (eelworms or roundworms) and, apart from protozoa, are the smallest of the soil mesofauna, being 0.5 to 1.0 mm in length. They are abundant in soil and litter and feed on roots, plant remains, bacteria and sometimes protozoa. Annelids are essentially enchytraeid worms (potworms) and lumbriscid worms (earth-worms). Enchytraeids are small (0.1–5.0 cm in length) and feed on algae, fungi, bacteria and soil organic matter. Earthworms are more important in the consumption of leaf litter than all other invertebrates together. Their numbers can exceed 800 m^{-3}. They feed exclusively on dead organic matter, and worm casts have a higher pH, exchange-able calcium, available phosphate and mineral nitrogen content than the surrounding soil. In this way, organic matter at the surface rapidly becomes incorporated throughout the upper part of the soil profile. Earthworms are rarely found in soils with pH values of less than 4.5. They are found in greatest abundance in mixed woodland and pasture soils and in least abundance in coniferous forest soils. Molluscs are not nearly so important as other meso-organisms and average about $200–300 \text{ kg h}^{-1}$. The main groups are slugs and snails, which feed on plants, fungi and faecal remains.

Micro-organisms contribute to the decay and breakdown of organic matter and to the break-up of mineral particles. These organisms are important to soil development. They release nutrients to plants and are essential to the operation of the nutrient cycles. It is the micro-organisms that are mostly responsible for the breakdown of leaf litter and other dead organic matter. The rapid decay and turnover of organic matter in tropical rainforest soils is due to the intense activity of micro-organisms. The operation of micro-organisms is accelerated under conditions of high temperature and humidity, whereas in cold environments the rate of decay of organic matter is slow. The main types of micro-organism are bacteria, actinomycetes, fungi, algae and protozoa.

Bacteria are as small as 1 μm in length and 0.2 μm in breadth. They generally live in the water films that exist around soil particles. They occur in enormous numbers ($1-4 \times 10^9 \, g^{-1}$) and in endless variety and possess the ability to decompose a variety of substances. **Heterotrophs** require complex organic molecules for growth. **Autotrophs** synthesise their cell constituents from simple inorganic molecules by harnessing the energy of sunlight (photosynthetic bacteria) or energy from the chemical reaction of inorganic compounds (chemoautotrophs). Bacteria may also be subdivided on the basis of their requirement for molecular oxygen. Aerobes require oxygen, whereas facultative anaerobes, while normally requiring oxygen, are able to adapt to oxygen-free conditions by using NO_3 and other inorganic compounds as electron acceptors in respiration. Obligate anaerobes grow only in the absence of oxygen.

Bacteria possess the ability to decompose a wide range of materials. *Pseudomonas* spp. can metabolise a range of chemicals including pesticides, *Nitrobacter* spp. obtain energy from the oxidation of nitrite to nitrate, and it has already been seen how *Rhizobium* spp. form nitrogen-fixing nodules.

Actinomycetes form colonies, similar to fungi. During their vegetative stage, actinomycetes form fine branching filaments (about 1 μm in diameter), but during reproduction, the filaments undergo fragmentation and sometimes form dense colonies. They are aerobic and less tolerant of soil acidity than fungi but are better able to decompose lignin and complex organic compounds than either fungi or bacteria. Their numbers in soils are similar to those of fungi, but their biomass is less.

Fungi occur as soil inhabitants, those residing in soil and feeding on dead organic matter, and as soil invaders, whose spores are normally deposited in the soil but which germinate and grow only when living tissue of a suitable host plant appears close by. They produce filamentous structures (hyphae) about 0.5 to 1.0 μm in diameter, which develop into a dense network or mycelium. Fungi are less numerous than bacteria ($1-4 \times 10^5$ organisms per gram), but their biomass can be a half to two-thirds the bacterial biomass. They grow better in acid soils (especially pH <5.5) and tolerate variations in soil moisture better than bacteria. In acid soils, they can be responsible for 60 to 80 per cent of organic matter decomposition. All fungi are heterotrophic and are most abundant in the litter layer and organic-rich surface horizon, and they play an important role in the development of soil structure (Molope *et al.*, 1987).

Algae are photosynthetic and are therefore confined to the soil surface. However, they may grow in the absence of light if simple organic solutes are provided. Cyanobacteria are the most important, especially the genus *Nostoc*, which can reduce atmospheric nitrogen and incorporate it into amino acids, thus making a contribution to the nitrogen status of the soil. Cyanobacteria prefer neutral to alkaline soils, whereas green algae are more common in acid soils. The abundance of algae can be as many as 3×10^6 organisms per gram.

Protozoa, the smallest of the soil animals (5–40 μm in length), are uni- or non-cellular organisms that live in water films. Nearly all protozoa prey on small organisms, bacteria, algae and nematodes, and are important in controlling the numbers of bacteria and fungi. They can be divided into amoebic forms, with silica or chitin sheets, and rotifers with better-differentiated cell structure.

Soil air

Soils can be said to 'breathe'. Soil animals, plant roots and most micro-organisms use oxygen and release carbon dioxide. Oxygen needs to move into the soil and carbon dioxide out of it to maintain biological activity. This movement is primarily by diffusion. Soil organisms use oxygen to oxidise carbohydrates. In temperate regions during the summer months, oxygen consumption by soils ranges from $4-20 \, \text{g m}^{-2} \text{d}^{-1}$ for arable soils to $10-20 \, \text{g m}^{-2} \text{d}^{-1}$ for soils under forest. In the humid tropics, rates vary between 8 and $50 \, \text{g m}^{-2} \text{d}^{-1}$. Plant carbohydrates move to the roots, where they are oxidised, releasing energy needed for plant growth. In the process, oxygen is taken in and carbon dioxide released at the root surface. The oxygen demand by soil organisms is large individually but small compared with that of plant roots and microbial populations. Microbes use carbohydrates in plant and animal residues in the breakdown of soil organic matter. Oxygen demand is greatest in the surface horizons, because this is where the maximum numbers of animals, micro-organisms and roots are found. The rate of oxygen use is known as the soil's respiration rate and is controlled by temperature, soil water content, soil organic matter content and nutrient supply. Oxygen demand is usually highest in summer months following rain. This means that oxygen concentrations are lowest during a wet period in summer and not during wet winter periods, because oxygen demand in winter is low. Conditions are said to be **anaerobic** if the supply of oxygen does not meet demand. The process by which oxygen is replenished and carbon dioxide removed is called soil aeration.

Soil air is an important constituent of soils in that it often determines the activity of other processes. The pressure of air in the pores compared with atmospheric pressure can also be significant. Soil air usually has the following composition: 79–80 per cent nitrogen, 15–20 per cent oxygen and 0.25–5 per cent carbon dioxide. There are usually traces of other gases that are by-products of microbial metabolism. Thus, as the above-ground atmosphere contains only 0.03 per cent carbon dioxide, there is a considerably enhanced amount of carbon dioxide in soils. This has been derived largely from the breakdown of organic matter and may dissolve in water to form a weak acid. This slight acidity will enhance weathering processes. The interaction between soil air and soil water will also be important.

Oxygen is replenished and carbon dioxide removed from the soil by two processes. Mass flow is a bulk movement that occurs as a result of changes in pressure or temperature. Expansion or contraction forces the gas in and out, and water passing through the soil can carry dissolved gases and may also push gas ahead of it. The most important process is diffusion, whereby gas molecules move in response to concentration gradients formed in the soil. Diffusion is easiest through continuous air-filled voids, particularly in the vertical direction. The rate of diffusion in air is about 10,000 times faster than in water.

Oxygen uptake and carbon dioxide production occur mostly in microbial cells and at plant roots. Figure 1.8 shows a diagrammatic representation of the complex system at an electron microscopic level. Gases must diffuse through thin films of water. Organisms appear to be able to continue respiration even when the oxygen concentration has fallen to very low levels. It has been found that if the oxygen concentration in air-filled pores is near to 21 per cent, oxygen supply can be maintained through an adjacent water-filled surface about 1 cm thick. This is known as the critical distance. Sandy, well-drained soils possess a good system of transmission pores, and active sites are not usually more than the critical distance from an air-filled pore. Under such conditions, there is unlikely to be a shortage of oxygen. In contrast, heavy-textured soils commonly exhibit poor aeration. Oxygen supply is restricted for three main reasons (Rowell, 1994):

1 Such soils are often in low-lying situations with a high **water table**. Thus, large volumes of soil will be saturated and anaerobic.
2 Heavy-textured soils often overlie clay subsoils

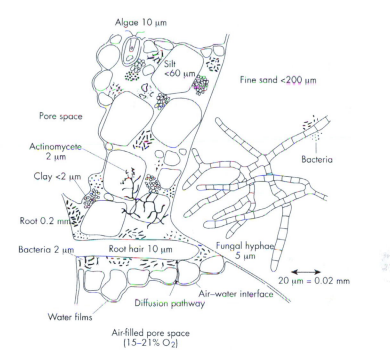

Figure 1.8 Mineral particles, water films and active sites of respiration.
Source: After Rowell (1994)

with low permeabilities. Water will be held at this level and a perched water table produced. Anaerobic conditions will again be created.

3 Some heavy-textured soils possess a poor structure with few transmission voids and will become saturated relatively easily.

The variation in aerobic and anaerobic zones in soils with different degrees of waterlogging is shown in Figure 1.9.

Soil water

Water is an essential part of the soil system and is necessary for plants and animals to survive. It is obtained either from precipitation or from ground-water and its composition varies considerably, even over short time periods. Variability is caused by the association between water, mineral and organic particles and plant roots. The chemical composition of precipitation also varies considerably. Water contains a number of dissolved solid and gaseous constituents, such as base **cations** (Ca^{2+}, Mg^{2+}, K^+, Na^+, NH_4^+) and a number of **anions** including chlorides. Water is held in small **pores** and in the necks between larger air-filled pores (Figure 1.10).

In its passage through soil, water transports solid mineral and organic particles as well as various dissolved components. This process is called **leaching**. Some of this material is removed from the soil completely, while some of it is redeposited lower down the profile. **Eluviation** is the movement of

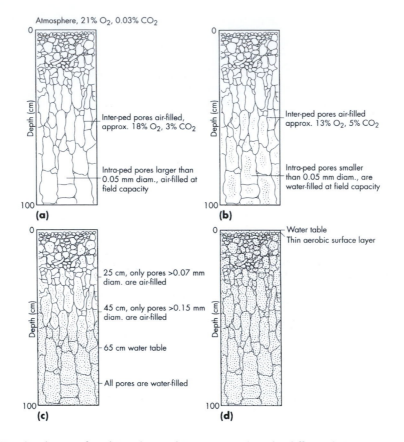

Figure 1.9 The distribution of aerobic and anaerobic zones in soils under different drainage regimes. (a) Well drained at field capacity and fully aerobic. (b) Imperfectly drained at field capacity and with anaerobic microsites. (c) Poorly drained at field capacity. (d) Waterlogged soil. The dotted areas are anaerobic.
Source: After Rowell (1994)

material down through the soil, resulting in a depletion in some horizons. **Illuviation** is the precipitation or accumulation of material, usually within the B horizon, after material has been leached from upper horizons. These processes are examined in greater detail in Chapter 3. It is also possible for water to move up the soil profile, under capillary action, and to transfer chemical salts to the surface horizon. This is especially prevalent in arid areas and produces distinctive saline soils. Water is important

in the weathering of the parent material and the production of new soil matter. Thus water creates, transforms and transports soil material.

The movement of water and the ability of the soil to hold water will largely depend on the soil's surface infiltration capacity and the rate at which water can percolate through the soil. Infiltration capacity (see Chapter 3) is determined by vegetation at the surface and soil properties in the upper horizons. Percolation depends on the nature and

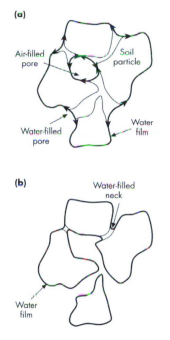

(a)

Air-filled pore

Soil particle

Water-filled pore

Water film

(b)

Water-filled neck

Water film

Figure 1.10 Distribution of water in soil.

abundance of soil pores and on soil permeability. If a soil is excessively porous and permeable, water will percolate rapidly through it and the soil will become dry. If a soil is impermeable, water will have great difficulty in passing through it and the soil water will probably be held tightly within the pores. In both instances, it will be difficult for plants to obtain the water they need. These properties are examined in greater detail in the next chapter.

SOILS AS LANDSCAPE SYSTEMS

It has already been stressed that soils do not exist in isolation but are organised within the landscape. A landscape system may be defined as any landscape unit in which the 'biosphere, toposphere, atmosphere, pedosphere and hydrosphere, together with

the biological, pedological and hydrological processes that create them, are seen as a unitary whole' (Huggett, 1995: p. 14). Soil landscape systems operate at a variety of scales. The smallest effective unit of soil is called a **pedon** (Simonson, 1978). A pedon consists of a small volume of soil starting at the surface and extending downwards to include the full set of horizons or complete soil profile down to parent material. Larger units composed of a number of similar contiguous pedons are called polypedons. The relationships between these are depicted in Figure 1.11. Polypedons will themselves form a mosaic of different soils within the landscape, different yet linked together as systems. The pattern of this mosaic will usually reflect the dominance of one controlling or formative factor over the others. This may be parent material, topography, climate, vegetation type or some other factor. The controlling factor or factors will depend on the size of the unit. Soils are often considered as two-dimensional bodies existing at some point on a slope transect. This could be a line system, and it is possible to subdivide this line system into a number of individual working subsystems, each with its own input, throughput and output. The detail of such a possible system on a valley side slope is shown in Box 1.2. Soil patterns at a variety of scales are examined in Chapter 7.

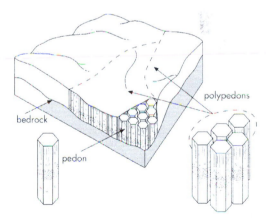

polypedons

bedrock

pedon

Figure 1.11 Arrangements of pedons and polypedons.

Box 1.2

TWO-DIMENSIONAL SOIL–SLOPE SYSTEM

This example demonstrates different degrees of complexity for the movement of water through the upper soil horizons on an individual valley side slope.

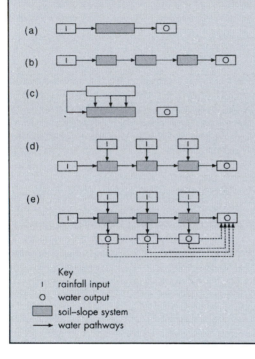

(a) This is the simplest case, and it is assumed that all water enters the soil system at the top of the slope and leaves it at the base.

(b) Slightly more realistic is where inputs and outputs are still considered *en bloc* but the soil system has been subdivided (perhaps into upper, middle and lower slope portions) with transfers between these portions.

(c) This scheme allows input to occur throughout the length of the slope, but the slope is not subdivided.

(d) This is a situation where inputs and through-puts are considered as separate, though related, components on different slope zones.

(e) This is a more realistic approach where inputs, throughputs and outputs are all composed of discrete units. In the water movement analogy being used, this allows infiltration and then percolation to lower soil horizons as well as throughflow, deep percolation and ground-water flow.

Key
I   rainfall input
O   water output
▢   soil–slope system
→   water pathways

SUMMARY

Soils, although difficult to define, can be described in terms of their basic mineral and organic constituents and their associated air and water properties. They act as open systems, with inputs, throughputs and outputs of materials and energy. They can also be regarded as one of the major geospheres (pedosphere). They also interact with the other main geospheres and are an essential part of a number of physical and chemical global cycles.

ESSAY QUESTIONS

1 **With the aid of specific examples, illustrate the way in which soils may be considered as open systems.**

2 **Justify the inclusion of the pedosphere in the main group of geospheres.**

3 **Describe and explain the operation of one of the main global cycles.**

FURTHER READING

Christopherson, R.W. (1992) *Geosystems: An Introduction to Physical Geography*, New York: Macmillan.

Huggett, R.J. (1995) *Geoecology*, London: Routledge.

Ollier, C.D. and Pain, C. (1996) *Regolith, Soils and Landforms*, Chichester: J. Wiley & Sons.

Stevenson, F.J. (1986) *Cycles of Soil: Carbon, Nitrogen, Phosphorus, Sulfur, Micro-nutrients*, New York: J. Wiley & Sons.

Strahler, A.H. and Strahler, A.N. (1992) *Modern Physical Geography*, 4th edn, New York: J. Wiley & Sons.

White, I.D., Mottershead, D.N. and Harrison, S.J. (1992) *Environmental Systems: An Introductory Text*, London: Chapman & Hall.

2

SOIL PROPERTIES

INTRODUCTION

Soil properties are specific attributes of the main soil constituents discussed in the previous chapter. They are capable of being measured either in the field or in the laboratory. The measurement of some of these properties is extremely difficult and beyond the scope of this book. However, many are capable of accurate assessment in the field with relatively simple techniques and procedures. A basic subdivision can be made between physical and chemical properties.

PHYSICAL PROPERTIES

Particle size

Soil is composed of particles of a variety of sizes. A general distinction is often made between boulders (several metres in diameter), pebbles and cobbles (several centimetres in diameter) and **sand, silt** and **clay** (< 2 mm in diameter). This last category is sometimes referred to as the fine fraction or **fine earth.** Within the fine earth fraction clay particles are less than 0.002 mm in diameter; silts are 0.002–0.06 mm in size (although the United States Department of Agriculture, USDA, uses 0.05 mm

as the upper limit of silt-size particles), and sands range from 0.05 to 2.0 mm. The various definitions of the fine earth fraction are shown in Figure 2.1. Larger particles, such as gravel and cobbles, are usually excluded from this analysis but can be taken into account by determining a soil's stoniness. Soil particles often occur as combinations or aggregates, and it is often the size of aggregates that determines a soil's behaviour (see section titled 'Soil structure'). Soil erodibility is a function of aggregate size and the ease with which aggregates are broken down by rain splash (see Chapter 8). The agricultural use of soils may be affected by aggregate size and stability. Particle size also affects the chemical characteristics of soils, because very small particles exhibit special properties.

Soil texture

It is extremely rare for soils to be composed of a single particle size class. Thus soil texture is defined in terms of combinations of sand, silt and clay using a soil texture triangle or pyramid (Figure 2.2a). These combinations result in eleven main textural classes. Again, it should be noted that the USDA scheme is slightly different, with twelve classes (Figure 2.2b). The difference is in the classification of loams. The USDA system has a loam category

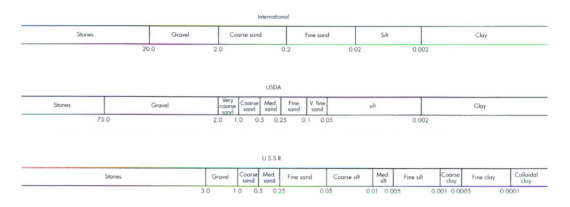

Figure 2.1 Three major schemes of particle size distribution (in millimetres).

that is similar to but not exactly the same as sandy silt loam, and a separate category for silt. A simplified triangular diagram is shown in Figure 2.3. A loam soil will possess approximately 40 per cent each of silt and sand and 20 per cent of clay-size particles. The sand, loamy sand, sandy loam and loam classes are divided into fine, medium and coarse according to the size of the sand fraction. Loams have the best combinations of physical and chemical properties for crop growth and cultivation. They allow easy drainage of excess water yet retain enough water for plant use. There also tends to be a good nutrient supply in loam soils. Coarse sands warm up quickly in summer and provide easy drainage, but they have a poor ability to hold water, are easily leached and possess low nutrient contents. Fine sands and silts are easy to cultivate but may be prone to erosion. Clays provide a good supply of nutrients for plant use but possess poor drainage, becoming waterlogged when wet and hard and cracked when dry. Hardsetting (see Chapter 8) decreases infiltration rates (see later in this chapter and also Chapter 3) and increases surface runoff. Very heavy clay soils can be improved for cultivation by increasing the organic content and perhaps adding substances such as gypsum. Stony soils are difficult to cultivate, are prone to drought and fail to retain nutrients.

Soil texture is determined in the laboratory by first removing the organic matter, usually by hydrogen peroxide, followed by sieving and by various sedimentation techniques for the finer fraction. The mineral soil may also need to be dispersed by shaking in the presence of sodium hexametaphosphate. It is also possible to estimate texture by more 'rough and ready' methods. As texture is related to the 'feel' of the soil, there is an appropriate key for a manual assessment (Table 2.1). Coarse sand grains are large enough to grate against each other and can be detected individually by sight and feel. Fine sands, when they form more than 10 per cent of the sample, can also be detected by feel. Silt makes the soil feel smooth and soapy and only very slightly sticky. Clay can mould and will impart a sticky feel to the soil. High organic matter contents tend to reduce the stickiness of clayey soils and make sandy soils feel more silty (see section titled 'Soil consistency').

Texture will affect processes operating within the soil and will affect chemical exchange because surface area per unit volume increases greatly as particle size decreases. Texture is an indication of the size of the particles and not necessarily their mineral composition. It is important to make the distinction between clay-size particles and clay minerals. Any mineral can be of a clay size. Quartz,

(a)

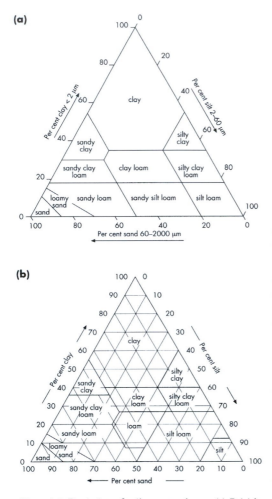

(b)

Figure 2.2 Depiction of soil texture classes. (a) British scheme. (b) USDA scheme.

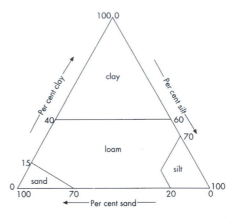

Figure 2.3 Simplified triangular diagram of soil texture.

a relatively inert mineral in soils, often occurs as particles within the clay size range. Most minerals, apart from carbonates and one or two others, are silicates, but clay minerals are so distinctive that they deserve special consideration (see section titled 'Soil mineralogy'). Clay-size particles tend to hold water because they are more compactible, which reduces the size of the soil pores. Soils with appreciable amounts of clay-size particles are porous but not very permeable. It can be argued that soil texture is the most important of all soil properties as it affects erodibility and water-holding capacity.

Particle surface texture

Much information concerning soil processes can be obtained by examining the surface texture of individual soil particles, using a scanning electron microscope. Specific surface features, such as etching pits, scratches and coatings, are often characteristic of specific genetic environments (Krinsley and Doornkamp, 1973; Whalley and McGreevy, 1983). Creemans *et al.* (1992) have identified the way in which the surface texture of feldspar minerals changed during the course of weathering. Alteration occurs at the grain surface/solution interface and preferentially at sites of excess energy such as points of crystal defect and twin planes, forming small etch pits, which enlarge, become trench-like in appearance and eventually coalesce to produce a honeycomb structure. These ideas have been used to produce a feldspar weathering index (Reed *et al.*, 1996). Care has to be taken to establish that surface texture characteristics are the result of the *in situ* effects of soil processes.

Table 2.1 Key for the assessment of soil texture

Grittiness	Smoothness	Stickiness and plasticity	Ball and thread formation	Texture
Non-gritty to slightly gritty	Not smooth	Extremely sticky and plastic	Extremely cohesive balls and long threads, which bend into rings easily	Clay
	Moderately smooth	Very sticky and plastic	Very cohesive balls and long threads, which bend into rings	Silty clay
		Moderately sticky and plastic	Moderately cohesive balls, forms threads that will not bend into rings	Silty clay loam
	Extremely smooth	Very slightly sticky and plastic	Moderately cohesive balls, forms threads with difficulty that have broken appearance	Silt
	Very smooth	Slightly sticky and plastic	Moderately cohesive balls, forms threads with great difficulty that have broken appearance	Silt loam
Slightly to moderately gritty	Slightly smooth	Moderately sticky and plastic	Very cohesive balls, forms threads that will bend into rings	Clay loam
Moderately gritty	Not smooth	Very sticky and plastic	Very cohesive balls, forms long threads that bend into rings with difficulty	Sandy clay
	Not smooth	Moderately sticky and plastic	Moderately cohesive balls, forms long threads that bend into rings with difficulty	Sandy clay loam
	Slightly smooth	Slightly sticky and plastic	Moderately cohesive balls, forms threads with great difficulty	Loam
Very gritty	Not smooth	Not sticky or plastic	Slightly cohesive balls, does not form threads	Sandy loam
Extremely gritty	Not smooth	Not sticky or plastic	Slightly cohesive balls, does not form threads	Loamy sand
		Not sticky or plastic	Non-cohesive balls that collapse easily	Sand

Source: Fitzpatrick (1983)

Particle shape

Particle shape is also often related to a specific genetic environment. Shape can be described in two or three dimensions. Common terms used include roundness, angularity and elongation, while in three dimensions terms such as sphericity and flatness and shapes such as sphere, disc, rod, cube and prism are described. There is a certain subjectivity in these terms, so a number of shape indices have been devised.

An interesting variation on this approach has been suggested by Green (1974), involving the long (p) and medium (q) dimensions of grains. It seems that it is possible to separate *in situ* soils from transported sediments by plotting the elongation function p/q against p for different grain sizes. Elongation functions for an *in situ* granite soil showed that its quartz grains became increasingly equant with increasing size. The coarser particles of an alluvial soil derived from granite debris were less equant and more elongated with increasing grain size, which also seems to be true of river deposits.

Soil mineralogy

Soil minerals occur in a variety of forms, but the most abundant group of minerals are **silicates** of one form or another. Silicate minerals consist of silicon–oxygen (Si–O) tetrahedra linked in a variety of ways. The tetrahedron building block consists of a central silicon ion (Si^{4+}) surrounded by four closely spaced oxygen ions (O^{2-}) (Figure 2.4). This means that with silicon being a quadrivalent cation and oxygen a divalent anion, every silica tetrahedron must have a negative charge of 4. The structure of silicate minerals can be explained by the way in which this negative charge is balanced. If tetrahedra are linked to produce a single chain, any tetrahedron will share two oxygens with neighbouring tetrahedra and two would remain unshared, leaving a negative charge of 2 (Figure 2.5a). If two chains are cross-linked, the external tetrahedra are linked as in single chains and the excess negative charge remains 2. However, the internal tetrahedra share three

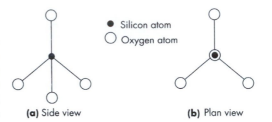

Figure 2.4 The basic building block of silicate minerals, the silicon–oxygen tetrahedron.

oxygens with neighbouring tetrahedra, so the total negative charge becomes 5, leaving an excess of only 1 (Figure 2.5b). In this double-chain structure, there are equal numbers of the two types of tetrahedra, so the overall excess negative charge is 1.5.

If three oxygens in any one tetrahedron are shared, a sheet structure results, and the overall charge on any one tetrahedron is reduced to 1. If all oxygens are shared with neighbouring tetrahedra, a three-dimensional network results and the positive charge on the silicon is balanced by the shared negative charge on the oxygens to produce an electrically neutral silicate. Thus these various interlinkings produce four different types of silicate: isolated tetrahedra (orthosilicates), chains (inosilicates), sheets (phyllosilicates) and three-dimensional networks (tektosilicates).

Apart from tektosilicates, the other silicate groups possess a residual negative charge that needs to be balanced by the additions of cations (see section titled 'Cation exchange'). Not all cations are equally suited to each of the silicate types. Size is a major factor, with, in general, the greater the valency the smaller the size. Univalent potassium is nearly the same size as oxygen, while bivalent aluminium is much smaller but not quite as small as quadrivalent silicon. In picometres (10^{-12} m), their sizes are 133, 140, 51 and 42 pm, respectively. Thus, potassium would require large gaps in the silicate framework to be accommodated, whereas magnesium and iron (66 and 64 pm, respectively) require only small gaps. In general, the greater number of Si–O–Si bonds the larger the gaps. Thus

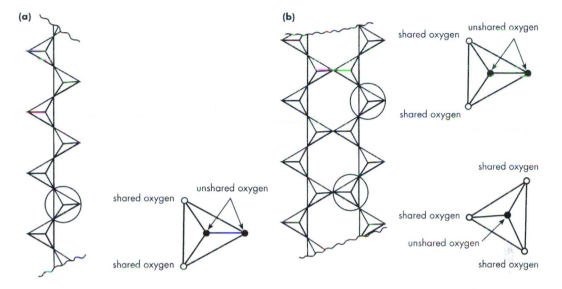

Figure 2.5 Tetrahedral linkage of silicate minerals. (a) Single chain. (b) Double chain.

orthosilicates are tightly packed, inosilicates and phyllosilicates less so, while tektosilicates (apart from quartz) have open structures. Thus, smaller cations, such as magnesium and iron, tend to be concentrated in orthosilicates and large cations, such as potassium, calcium (99 pm) and sodium (97 pm), are concentrated in tektosilicates.

Sometimes there is a substitution of one ion for another within the mineral structure. This is known as **isomorphic substitution**. One of the most common substitutions is the replacement of some Si^{4+} ions by Al^{3+} ions. Substitution may also alter the overall charge balance, frequently resulting in an excess negative charge. The substitution of Si^{4+} with Al^{3+} is one example. Neutrality will then be restored by the addition of cations, as previously explained. Isomorphic substitutions are related to environmental conditions, and identification of the specific nature of the substitutions may therefore tell us something about the environment at the time of formation.

These mechanisms produce a wide variety of mineral types. The tektosilicate group includes some of the most important minerals, such as quartz and feldspars. Quartz is made up of silicon tetrahedra in which the silicon ion fits between four oxygen ions, covalently bonding them by sharing one electron with each. The structure is compact and chemically strong, making quartz very resistant to weathering. Feldspars contain significant amounts of Al^{3+} ions originating from isomorphous substitution of Si^{4+} ions, plus base cations, usually Ca^{2+}, Na^+ and K^+. Feldspars can be divided into plagioclase and orthoclase varieties. Plagioclase feldspars vary from the high-temperature calcic plagioclase anorthite to the low-temperature sodic variety albite. As more substitutions occur, the structure becomes chemically weaker. Thus anorthite, in which aluminium replaces every other silica, is less resistant to weathering than albite, where aluminium replaces only every fourth silica ion. Orthoclase and microcline are the commonest members of the potassium-rich feldspar group.

Inosilicates possess chain structures, which create well-defined prismatic cleavages. The chains are linked together by a variety of cations. Minerals in this group are very reactive and are more easily weathered than tektosilicates. Pyroxenes are typical single-chain inosilicates, and amphiboles are double-chain inosilicates.

Orthosilicates and ring silicates display a wide variety of structures, but their occurrence in soils is relatively minor. They can be divided into two groups – nesosilicates and sorosilicates. In the first group, the Si–O tetrahedra occur as separate units, with no shared O^{2-} ions, and are linked by metallic cations. The most common examples are olivine, the garnets and zircon. In sorosilicates, the tetrahedra form separate groups in which they share one or more of their O^{2-} ions. If the group is formed by sharing O^{2-} ions with more than two Si–O tetrahedra, a ring structure results. Sorosilicates do not include important soil minerals. The most common are epidote, cordierite and beryl.

The sheet silicates or phyllosilicates play perhaps the most important role in soils. Because they possess a sheet structure, many have a well-developed cleavage, which aids the weathering process. The micas are good examples of this. Many sheet silicate minerals occur in the smallest grain-size categories and are known as **clay minerals**. All clay minerals are made up of combinations of two main sheet structures; a tetrahedral Si–O (siloxane) sheet and an octohedral sheet dominated by closely packed hydroxyl (OH^-) ions. This last sheet can be divided into an Al–OH (**gibbsite**) sheet, in which Al^{3+} ions are surrounded by the OH^- ions, and a Mg–OH (brucite) sheet, which contains Mg^{2+} ions rather than Al^{3+} ions.

Clay minerals are classified on the basis of the ratio, in a unit layer, of these sheets, the spacing between unit layers and their interlayer components. Perhaps the most important distinction is between 1:1-type clay minerals and 2:1-type minerals. The 1:1-type minerals are composed of one tetrahedral siloxane alternating with one octohedral sheet. **Kaolinite** (Figure 2.6c) is the commonest 1:1 clay mineral with one siloxane sheet to one gibbsite sheet and an interlayer spacing of about 7 Å (0.7 nm). Halloysite and allophane are

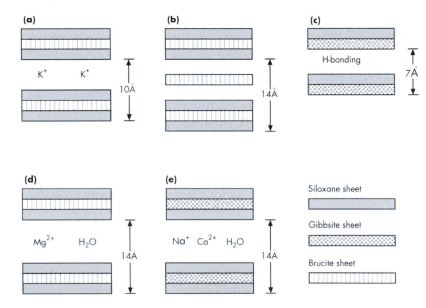

Figure 2.6 Representation of clay mineral structure. (a) Biotite. (b) Chlorite. (c) Kaolinite. (d) Vermiculite. (e) Montmorillonite.

also 1:1 minerals. Halloysite is similar to kaolinite but with curved crystal layers. Allophane is a gel-like hydrated aluminosilicate mineral.

There are many 2:1 clay minerals. Both main micas (biotite and muscovite) are 2:1 minerals, with two siloxane sheets to one brucite sheet for biotite (Figure 2.6a) and a gibbsite sheet for muscovite. In biotite, interlayer spacing is about 10 Å (1.0 nm) and K^+ is the dominant interlayer component. Other common 2:1 minerals are **illite**, which is very similar in composition to mica, and vermiculite (Figure 2.6d), which has interlayer positions occupied by Mg^{2+} ions and water molecules and a basal spacing of about 14 Å (1.4 nm). Chlorite (Figure 2.6b) has a 2:1:1 layer structure (two siloxane sheets to one brucite forming a biotite mica layer, to one additional brucite sheet occupying the interlayer position). Because of their property of shrinking and swelling on drying and wetting, there is a group of clay minerals known as swelling clays (smectites). **Montmorillonite** (Figure 2.6e), although difficult to identify, is a 2:1-layer mineral with interlayer positions dominated by Na^+ and Ca^{2+} ions and water molecules. Its basal spacing is approximately 14 Å (1.4 nm) but will depend on whether it is in a 'shrink' or 'swell' state.

A number of mixed-layer minerals can also occur. These result from the interlayering of more than one sheet silicate mineral. They usually exist as intergrades when one mineral is being transformed into another during weathering (see Chapter 3). Examples include illite–montmorillonite, vermiculite–chlorite, montmorillonite–chlorite,

illite–vermiculite, mica–vermiculite and kaolinite–montmorillonite. The properties of the more important clay minerals are summarised in Table 2.2.

There are also a few non-silicate minerals in soils such as free oxides and hydroxides of iron, aluminium and magnesium. Iron minerals include goethite, maghemite, ferrihydrite, haematite, lepidocrocite and magnetite. Other non-silicate minerals are carbonates, such as those of calcium and magnesium, and anatase (TiO_2) and amorphous silica, which are often found in volcanic soils. Carbonates may not dominate a soil in terms of volume or weight, but they can exert a considerable influence on soil processes. FitzPatrick (1986) has listed the following important properties of carbonates:

- they are easily soluble in water and therefore can be lost from or redistributed within the soil;
- even small amounts of carbonate may raise the pH value of the soil and help to sustain a high level of biological activity;
- carbonates, especially calcium carbonate, are the first substances to start accumulating when the climate becomes arid; and
- both calcium and magnesium are essential plant nutrients.

Soil density

It is important to make the distinction between the density of the individual soil particles and the **bulk density** of a specific part of the soil. The mineral

Table 2.2 Some properties of the more commonly occurring clay minerals

Clay	Layer structure	Layer thickness (nm)	Surface area ($m^2 g^{-1}$)	Swelling properties
Kaolinite	1:1	0.7	10	none
Illite	2:1	1.0	20	none or very little
Smectite	2:1	1.0	800	extensive
Vermiculite	2:1	1.4	400	limited
Chlorite	2:2	1.4	10	none

Source: After Rowell (1994)

grains in many soils, because of their abundance in rocks and their relative resistance to weathering, are mostly quartz and feldspar. This is especially true of the sand content, and a density of 2.65 g cm^{-3} is a realistic average figure. The density of individual minerals varies from 5.2 g cm^{-3} for magnetite to about 2.60 g cm^{-3} for feldspars. Organic matter has a density of about 0.9 g cm^{-3}. Bulk density is defined as the weight of soil per unit volume (g cm^{-3}). This can be a wet bulk density, which includes the water content, but it is usually expressed on an oven-dry basis. Bulk density reflects porosity and the proportion of organic matter. The bulk densities of most soils vary between 1.0 and 2.0 but can be 0.55 for some soils developed on volcanic ash (see Table 2.3). Soils with high bulk densities inhibit water movement and root penetration. Trampling and loading of the soil surface will increase bulk density and lead to more surface runoff. Ploughing continuously to the same depth may produce a compacted plough pan of increased bulk density. It is possible to estimate the depth to which water will wet a soil by using moisture content and bulk density values.

Soil porosity and permeability

Pores are the part of the soil occupied by soil air or soil water. Pores may be discrete or they may form continuous passages through which movement may take place. A distinction is often made between macropores, which may be up to several centimetres in diameter, and micropores, which may be less than 1 μm in diameter. A diameter of 60 μm (0.06 mm) is often taken as the size boundary between micropores and macropores. Pores may be formed in a number of ways. The formation of pores is often the first stage in the formation of peds. Pores may be created by wetting and drying, heating and contraction, and freezing and thawing. Pores may also be formed by the solution of minerals. Macropores are often created by faunal activity and the shrinking of clay-rich soils. They are important because they allow rapid water movement, sometimes known as water bypassing, and are critical in leaching and in the movement of fertilisers and pesticides in soils (Bouma et al., 1977; Flury et al., 1994; Nortcliff et al., 1994; see Chapter 8).

Macropores allow rapid drainage of water after irrigation or heavy rainfall, and when these pores are emptied drainage becomes very poor. Pores are sometimes classified according to function. Transmission pores (equivalent to macropores) allow movement of water after saturation and allow movement of oxygen and carbon dioxide when the soil is at field capacity if pores are continuous in a vertical direction. Macropores also allow root penetration. Storage micropores (50–0.2 μm) store water, which is then available for plant use, and residual micropores (< 0.2 μm) hold water so strongly that it is not available for plant use. Three critical conditions have been identified (Rowell, 1994): when transmission pore volume is below about 0.1 cm^3 cm^{-3}, drainage problems may occur; with storage pore volumes below about 0.15 cm^3 cm^{-3}, restricted water availability may

Table 2.3 Some typical bulk densities and porosities of soils

	Dry bulk density (g cm^{-3})	Porosity (cm^3 cm^{-3})
Cultivated mineral soils:		
medium to heavy texture	0.8–1.4	0.69–0.46
light texture	1.4–1.7	0.46–0.35
Subsoils	1.5–1.8	0.43–0.32
A horizons	0.8–1.2	0.67–0.50

Source: After Rowell (1994)

occur; if residual pore volumes are above about $0.2\,cm^3\,cm^{-3}$, the soil may possess difficult mechanical properties, being plastic and sticky when wet and hard when dry.

Porosity is an expression of the volume of pores to the total volume of the soil. It can be determined from particle and bulk density by:

$$Porosity\ (\%) = 1 - \frac{bulk\ density}{particle\ density} \times 100$$

It can also be determined from an undisturbed soil core of known volume (V) by the following:

$$Porosity\ (\%) = \frac{W_s - W_d}{V} \times 100$$

where W_s is the weight of the core saturated with water and W_d is the dry weight. Some typical porosity values are shown in Table 2.3.

Permeability values express the ease with which water can pass through the soil. It conveys information about the degree of interconnection between pores. Although porosity and permeability are related, the relationship can be quite complicated. Pores need to be of a sufficient size and to be interconnected for flow to occur. Clays have high porosities but low permeabilities, whereas a silty sand with a low porosity may have a high permeability rate.

Soil structure

Soil structure refers to the way in which individual particles are aggregated or joined together. It may be defined with reference to three characteristics:

1 the size, shape and arrangement of the particles and aggregates;
2 the size, shape and arrangement of the voids; and
3 the combination of voids and aggregates into various types of structure.

It is macrostructure that is usually defined in these terms. Microstructure can only be determined under a high-powered microscope.

Soil particles may be aggregated in a number of ways depending on particle size and type, the amount and nature of organic compounds, and climatic conditions. However, the origin of some structures is uncertain. Soil structure will affect permeability, root penetration, infiltration and percolation rates, and soil erodibility. Individual aggregates, called **peds**, are arranged and classified into a number of basic forms (Figure 2.7). The main forms are granular, blocky, prismatic, columnar and platy. A *granular* form is composed of spheroidally shaped aggregates with faces that do not accommodate adjoining ped surfaces. The term 'crumb structure' is applied to a porous granular structure. Granular structure and crumb structure are sometimes classed together as spheroidal structure. Granular and crumb structures are most common in A horizons, where the presence of roots and the action of soil organisms have helped their development. The particles appear to be bound together by organic **colloids**, which are small gelatin-like particles of minerals and organic matter that can attract and hold ions. Clay and iron and aluminium hydroxides may create some of the binding. Periodic drying out seems to help to form more stable aggregates.

Multivalent cations, especially Ca^{2+}, Mg^{2+} and Al^{3+}, are able to form attachments with more than one colloidal particle. This process is known as cation bridging. Silicate particles may be joined together by the attraction between positive charges at the broken edges of silicate sheets and the negative charges on the faces of similar particles. If soils possess a significant negative charge, anion bridging may occur. Inter-aggregate attraction can also be the result of various organic compounds, fungal hyphae, plant roots and organic polymers (Churchman and Tate, 1987; Haynes and Swift, 1990). At what is known as the domain level (up to $5\,\mu m$), electrostatic forces predominate and are resistant to change. At the microaggregate level (up to $250\,\mu m$), binding forces are still reasonably persistent, depending on organic content (Piccolo and Mbagwu, 1990). Binding forces at the macroaggregate level, especially those influenced by the

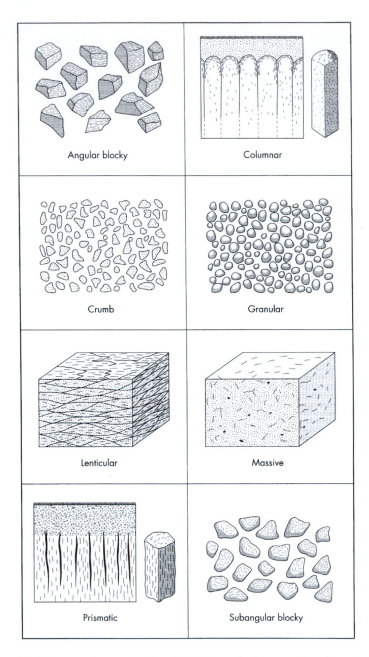

Figure 2.7 Diagrammatic representation of main types of soil structure (not drawn to scale).

nature of the vegetation and the development of root networks, are more transient.

'*Blocky*' is the name given to a structure composed of approximately equidimensional blocks possessing planar faces that are accommodated to adjoining ped faces. The intersections of faces are sharp in angular blocky structure and more rounded with subangular blocky structure. There is some doubt concerning the way in which blocky structure develops. One suggestion is that the faces are intersecting shear planes developed during swelling and shrinkage associated with soil moisture changes. In *prismatic* structure, particles are arranged about a vertical axis and each ped is bounded by planar, vertical faces that accommodate adjoining faces. In prismatic structure the peds possess a flat top, whereas in columnar structure peds have rounded tops. The vertical faces develop by shrinkage during dehydration and may possess small grooves or slickensides, indicating repeated swelling and shrinkage. The rounded tops in columnar structure may be due to greater amounts of upward swelling in the centres of columns during the wetting process. In platy or lenticular structure, particles are arranged about a horizontal axis. The origin of this structure is again uncertain, but it will be helped by particle-size orientation inherited from the parent material or by depositional processes. Freeze–thaw processes in the soil may aid its formation. It may also be related to layering created by the precipitation of minerals such as carbonate, silica and iron. It also develops as a result of compaction.

Water is transmitted between peds as well as between particles. A well-developed crumb structure, with numerous well-connected voids, will permit water movement in all directions. Blocky structure possesses many voids, but regular packing keeps void sizes small and water movement is again possible in all directions. Prismatic structures possess large peds with well-defined voids having a dominant downward orientation, and water movement is mainly in a vertical direction. *Platy* structures have large peds will ill-defined voids having a dominant lateral orientation, and water movement will be forced in a lateral direction. It is

quite common to find that different horizons in the same soil possess different structures. A horizons usually have a crumb structure, B horizons perhaps prismatic and C horizons blocky.

Structure can also be classified in terms of class, reflecting size, and grade, which varies according to distinctness and durability. Grade is defined in the following terms. In a structureless or apedal soil there is no observable aggregation or definite arrangement of lines of weakness. Such soils are massive if coherent and single-grain if non-coherent. In a weakly developed structure there are poorly formed, indistinct peds, which readily break down on disturbance. If the structure is moderately developed, the peds are well formed, distinct and moderately durable. The soil breaks down into a mixture of entire peds. A strongly developed soil structure possesses durable peds, which adhere weakly to one another. The soil consists of entire peds. Drying will increase the structure grade of clay soils (see also section titled 'Soil fabric').

Soil micromorphology

The micromorphology of soils comprises three components: plasma, skeleton, and voids. The plasma is the colloidal ($< 2 \, \mu m$) part and the soluble materials not bound up in the skeleton grains; it is the part that can be moved, reorganised and/or concentrated during soil formation. Skeleton grains include detrital mineral fragments, and secondary crystalline and amorphous bodies, which are not usually translocated, concentrated or reorganised during soil formation. Voids are the spaces within plasma and between plasma and skeleton grains. Plasma often occurs as coatings (**cutans**) on mineral grains. Clay coatings are known as **argillans**, and sand/silt coatings are **skeletans**. There may also be coatings of oxides, hydroxides and oxyhydroxides of iron and aluminium.

Clay coatings are formed by the progressive deposition of clay-size particles. Particles are deposited tangential or parallel to the surface, so they build up a series of layers. Such coatings are fragile and are sometimes very transient. It was originally thought that clay coatings formed by vertical clay translocation and accumulation, but it is now known that they can form in many ways. It is clear that many coatings are formed *in situ*, maybe as a result of rearrangement within wet horizons or where weathering is so rapid that clays so formed are often deposited where they have formed. Coatings of oxides and hydroxides of iron and aluminium are often associated with organic matter, especially in the middle horizons of podzols (see Chapter 3). Silt coatings are usually the result of detrital silt. Calcite coatings may form by the slow growth of calcite crystals from percolating soil solutions. If these coatings develop sufficiently, they may join together to form massive carbonate. In some environments, such as semi-arid areas, the soil solution may contain large amounts of silica, which can be deposited as coatings in the form of chalcedony or opal.

Plasma may be concentrated as hard **pans** and **plinthite**. These are usually of iron or silica and are generally cemented, but **fragipans**, areas of high bulk density but which are not cemented, also occur. Fragipans appear to have been compacted by ground ice action as well as aggregation by clay minerals, iron and aluminium compounds, or silica. Indurated deposits of calcium carbonate are known as petrocalcic horizons. Petrogypsic horizons are indurated layers of gypsum-rich material.

There are also more substantial indurated layers, known as duripans, or **duricrusts** when they occur on the surface. One of the most quoted definitions of duricrust is that of Bates and Jackson (1987), namely 'a hard crust on the surface of, or layer in the upper horizons of, a soil in a semi-arid climate. It is formed by the accumulation of soluble minerals deposited by mineral-bearing waters that move upwards by capillary action and evaporate during the dry season.' As will be noted in Chapter 3, there is some doubt about the role of upward moving water. Hard pans and plinthite do not possess the concentration of specific elements that characterise duricrusts. More specific names are used to designate their composition: thus *ferricrete* is rich in iron, *calcrete* is rich in calcium carbonate, *silcrete* is rich in

silica, *gypcrete* is rich in gypsum, and *manganocrete* is rich in manganese.

The nature and arrangement of voids often provides useful information about past and present soil processes. Many arctic and alpine soils, usually those associated with patterned ground, possess bubble-like pores or vesicles (van Vliet-Lanoe, 1985). Some vesicles may be caused by needle-ice crystals, but other origins are possible. Wetting processes and the expulsion of air during freezing of wet soils have been suggested as mechanisms. In a series of experiments, Harris (1983) demonstrated that vesicles formed by thixotropic behaviour of soils during thaw consolidation. The presence of vesicles in soils in patterned ground suggests that liquefaction is an important process in the development of such patterns.

Soil fabric

The way in which the solid particles are arranged with respect to one another is called the matrix fabric. A distinction can be made between incoherent and coherent fabric (Paton *et al.*, 1995). Soil material that lacks coherence behaves as if it were single-grained. Coherent materials have a stable relationship between solid particles and voids. This is what is meant by 'matrix fabric'. Coherent fabric can be divided into grain support fabric, where sand grains are in contact with one another, and plasma support fabric, where sufficient clay particles are present to surround the sand grains (Figure 2.8). Grain support fabrics tend to be loamy sands and sandy loams, and plasma support fabrics are usually clay loams or even heavier soils.

Each of these fabrics can be further differentiated into porous or dense depending on the closeness of packing of the individual grains. A fourth differentiating characteristic, noted by Paton *et al.* (*ibid.*), is the degree to which fabrics have been inherited from bedrock or have been formed within the soil. This may be one of the fundamental distinctions between the solum and saprolite, or topsoil and subsoil. It will be seen in Chapter 4 that many rock weathering profiles retain much of the original

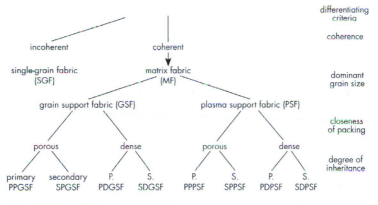

Figure 2.8 Classification of soil fabric.
Source: After Humphreys (1985)

structure of the rock. The soil proper has had its fabric altered by mesofaunal activity. Single-grain fabric may occur in the very top of the soil as a result of **rain splash** activity. Thus, in general, subsoils possess a primary fabric, often dense and plasma supported, whereas topsoils possess secondary, grain support fabrics.

Soil consistency

Soil consistency is an indication as to how soil material behaves when it is manipulated in its natural state. Terms used to describe consistency include 'brittle', 'plastic', 'friable', 'compact', 'loose', 'soapy', 'firm', 'sticky', 'tenacious' and 'thixotropic'. A brittle soil ruptures suddenly, whereas a plastic soil can be moulded when moist. Friable material is weakly coherent and easily crushed by gentle pressure. Friable material will become coherent again when pressed together. Compact soil is firm and moderately coherent with close packing. Loose soil is the same as soil with a single-grain fabric. Soapy material is sticky and plastic with a distinctive soapy feeling. It is usually the result of a high content of exchangeable sodium. Stickiness, the ability to adhere or stick to other objects, occurs only in some soils. A tenacious soil is plastic and requires a considerable pressure to mould

it. Thixotropic material becomes very wet when manipulated continuously between the fingers.

Consistency is influenced by texture and clay mineralogy. Butler (1955) has devised a simple test to estimate consistency (Box 2.1). A 3 cm cube of soil is extracted with as little disturbance as possible and is then squeezed between the palms of the hand until disruption occurs. The material is then subjected to a shearing stress by slowly rubbing the hands together with the same pressure as that required to cause the initial disruption. The degree of plasticity or brittleness is assessed by the pressure required for disruption, the nature of the fragments produced after disruption and how these fragments alter with the shearing stress. Grain support fabrics tend to be brittle, while plasma support fabrics are usually plastic. The moisture content of the soil will also affect its behaviour.

A rather more precise set of tests has been devised to test engineering materials, but they can be performed on soils. These tests define what are known as consistency limits. They are also known as Atterberg limits, after the Swedish agricultural scientist who devised the tests. The upper or **liquid limit** (LL) is the water content at which the soil, in a remoulded state, ceases to be plastic and may be considered a fluid. It is the minimum water content at which the soil will flow under its own weight.

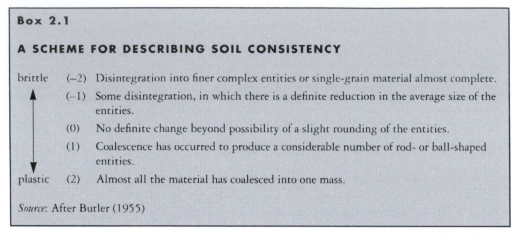

Box 2.1

A SCHEME FOR DESCRIBING SOIL CONSISTENCY

brittle (−2) Disintegration into finer complex entities or single-grain material almost complete.

(−1) Some disintegration, in which there is a definite reduction in the average size of the entities.

(0) No definite change beyond possibility of a slight rounding of the entities.

(1) Coalescence has occurred to produce a considerable number of rod- or ball-shaped entities.

plastic (2) Almost all the material has coalesced into one mass.

Source: After Butler (1955)

The **plastic limit** (PL) is the water content at which the soil ceases to be plastic and becomes friable or brittle. The shrinkage limit is the water content at which further loss of moisture does not cause a decrease in the volume of the soil. A measure of the range of water content over which a soil is plastic is known as the **plasticity index** (PI):

$$PI = LL - PL$$

Related to consistency is the shrink–swell potential of a soil, which can be determined from the coefficient of linear extensibility (COLE), the volume change on wetting and drying:

$$COLE = \frac{L_m - L_d}{L_d}$$

where L_m and L_d are the wet/dry sample lengths.

Soil temperature

Temperature will affect soils in several ways. Chemical reactions are enhanced at higher temperatures. Temperature influences the nature and productivity of plant growth, and the rate of organic matter decay. High temperatures, in the absence of water, will lead to the upward movement of capillary water, bringing with it dissolved chemical salts, which may then be precipitated at the surface. Extremely low temperatures will inhibit organic matter decay and will reduce the presence of living organisms. In high latitudes, the almost complete absence of vascular plants, low temperatures and low precipitation levels inhibit the movement of organic matter into the soil. Water freezing in the soil will affect soil structure and micromorphology. Soil heaving and churning processes (**cryoturbation**) can produce involutions and festoons within the soil body. Migration of water during soil freezing may cause dehydration of the soil immediately below the freezing front. Also, frost-heave pressures lead to compaction of unfrozen soil, and dense platy peds may be formed by compaction. Soil temperatures may be affected by soil characteristics such as colour, as dark soils absorb more heat than light-coloured soils. The percentage of the incident global radiation reflected by a surface is known as its albedo (see later under soil colour). The albedo of different plant communities also varies, which will affect soil temperatures. A scheme for defining soil temperature regimes is shown in Table 2.4.

Soil moisture

Water is an important constituent of soils. It may be freely available, or it may be held by adhesive forces between water molecules and organic and

Table 2.4 Soil temperature regimes

Name of regime	Mean annual soil temperature °C (T)	Difference between mean temperature (°C) of warm season and cold season
Pergelic	$T < 0°$	–
Cryic	$0° < T < 8°$	–
Frigid	$T < 8°$	$> 5°$
Mesic	$8° < T < 15°$	$> 5°$
Thermic	$15° < T < 22°$	$> 5°$
Hyperthermic	$T < 22°$	$> 5°$

Source: Based on Soil Survey Staff, *Soil Taxonomy* (1975, 1992)

inorganic particles and by cohesive forces between adjacent water molecules. Moisture content is usually expressed as a percentage of the oven-dry soil weight, but it can also be expressed on a volumetric basis using soil cores of known volume. Moisture content can also be related to its availability to plants. Water enters the soil by a process called **infiltration. Infiltration capacity**, which is the maximum rate of water movement into the soil, is controlled by many factors, soil structure and texture probably being the most important. Infiltration as a process is examined in Chapter 3.

Water may occur as gravitational water, capillary water or hygroscopic water (Figure 2.9). The distinction between these types is usually on the basis of soil tension. Hygroscopic water is adsorbed onto colloidal particle surfaces in thin films only a few molecules thick at high tensions (30 to <1,000 atmospheres). Capillary water is held in small micropores at tensions of 0.05 to 30 atmospheres. Water held at tensions below 0.05 atmospheres occurs in macropores and is gravitational water. The combined effect of adsorption forces and capillarity is known as **matric suction**.

Gravitational water passes freely down through the larger pores in the soil after precipitation in response to the pull of gravity. Water moves downwards into a homogeneous dry soil at a constant velocity to form a **transmission zone** behind a narrow wetting zone and a well-defined wetting front. However, because of variability in soil properties, actual downward water movement is more erratic. The depth to which soils are wetted is related to soil moisture retention. Studies have shown that the water-holding capacity of a 10 cm layer of clay is 1.4 cm; that for a silt loam is 1.7 cm and that for a sandy loam is 1 cm. Thus for a given rainfall amount sandy soils are wetted to greater depths than more heavily textured soils. If there are impermeable layers within the soil, the water may move laterally in a downslope direction as soil **throughflow**. It is now recognised that the downslope movement of water within soil layers is more important than overland flow, especially in most humid areas, where the supply of humus and the effects of microfauna and flora create an open soil structure. Even when there are no obvious discontinuities in the soil, subtle differences in the proportions of solid matter and water occur that may be sufficient to start the lateral movement of water.

Gravitational water may completely fill the pore spaces and is the water that is the first to be depleted in dry periods. As mentioned previously, **capillary water** is water that is held by surface tension on and within soil particles and in parts of the pore spaces. Capillarity is most effective in loam-textured soils. Capillary water may move downwards, upwards or laterally depending on climatic conditions. Capillary water is used extensively by plants, and it is the availability of such water during dry periods that allows many plants to survive. The upward

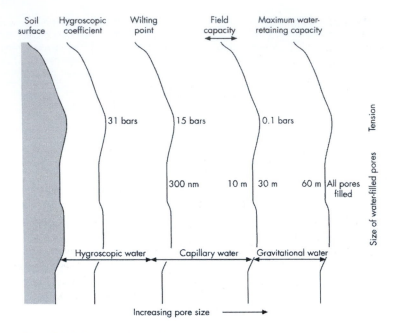

Figure 2.9 Soil–water relationships.
Source: Based on FitzPatrick (1983)

movement of capillary water is especially significant under such conditions. As capillary water is held in the soil by surface tension, it is removed from the soil only after long dry periods. When the forces holding water to soil particles are equal to the forces of downward gravitational pull, the soil is said to be at **field capacity**. At this point, water can be removed easily from the soil by evaporation and transpiration through the vegetation. As more water is removed from the soil, water films become thinner and are held more strongly to particles. Eventually, this water is held so strongly that roots cannot extract it. The water content at which this occurs is called the permanent wilting point. Water retention is largely determined by organic matter and clay contents. Water movement is influenced by bulk density, porosity and permeability.

Water that is held as microscopically thin films around individual particles is called **hygroscopic water**. This is held so tightly onto the soil particles that it is of little use to plants and is the last water to be removed from soils. It is probably not even removed during oven drying and does not figure in most soil moisture determinations. It occurs mostly on the surface of clay minerals, where negative charges attract positive charges on the ends of water molecules.

Soils may be described as excessively drained, well drained, imperfectly drained, poorly drained, very poorly drained or flooded, depending on their moisture contents. Excessively drained soils are usually too dry to support adequate plant growth. Well-drained soils usually contain enough moisture to support plant growth. Such soils usually possess bright colours and a range of textures. Imperfectly drained soils usually experience long periods when they are moist and short periods when anaerobic conditions prevail. This may produce mottles in the soil. Poorly drained soils are anaerobic for long periods of time and possess grey, olive and blue colours, indicating reducing conditions. Very poorly drained soils are saturated continuously and are uniformly blue, olive or grey. Flooded soils are periodically inundated but dry out during intervening periods.

Soil colour

Colour is often the most obvious soil characteristic and may indicate some of the processes that are or have been operating within the soil. Colour is usually determined by the fine material or matrix. Variations in soil colour are principally the result of variations in parent material, minerals, organic material and moisture content. Of these factors, moisture content is often dominant in determining colour. There is a general gradation from well drained to waterlogged, with colour changing from reds, browns and yellows to greens and blue-greys in response to changing oxidation/reduction balance.

Soils tend to be darker near the surface because of the accumulation of organic matter. The darkness and intensity of the colour reflects the nature of the organic matter and the stage in its breakdown. There is a tendency for organic matter to become darker in colour with increasing humification. Dark soils also develop on dark-coloured volcanic ash. Yellow-brown to red colours generally indicate the presence of iron (ferric) oxides, usually goethite and haematite, and are characteristic of the middle and lower parts of some tropical soils. The mineral responsible for most of the inorganic colouring in aerobic soils is goethite, with colours ranging from reddish brown to yellow. Some red soils, called terra rossa, are characteristic of certain limestone parent materials. Greyish colours are usually due to ferrous iron compounds formed under reducing (gleying) conditions. White or light grey colours may represent depositions of calcium carbonate or efflorescences of salts. Alternatively, they may indicate a horizon from which leaching has removed oxides or hydroxides of iron and aluminium. Also,

light-coloured topsoils tend to occur in high latitudes in humid environments, reflecting the paucity of organic matter. While colour itself may indicate the operation of certain processes, the intensity of the colour can be misleading. It takes very little iron to make a soil very red or orange. Intensity may be related simply to grain size in that it takes more colour to 'smear' the surface of larger particles.

As noted earlier, soil colour determines soil albedo. Pale sand has an albedo value ranging from 25 to 45, whereas values for a dark soil may range from 5 to 15. Deforestation and erosion of topsoil in semi-arid areas have exposed soils with relatively high albedo values, and it has been suggested that this has affected evaporation rates, leading to a decrease in precipitation.

Colour may occur in a variety of patterns such as spotted, streaked, speckled, marbled, and mottled or blotched (FitzPatrick, 1986). Individual spots of one colour in a uniform matrix would be classed as spotted, whereas streaked would be elongated areas of one colour in a uniform matrix. Speckled is characterised by small patches of organic and/or mineral matter fairly uniformly distributed in a matrix. With marbled patterns, colour differences are distinct but merge into one another in an irregular manner. Mottled or blotched refer to irregular patterns of two or more colours with sharp boundaries between the different areas. There are likely to be relationships between the presence of mottles and the degree of wetness (Veneman et al., 1976). Mottles may be good indications of contemporary pedological processes and are commonly used in field descriptions of soil profiles (e.g. Northcote, 1979; Soil Survey Staff, 1975; McDonald et al., 1990).

Colour is a useful characteristic, and there are standard keys for colour chroma and hue, which ensure a relatively objective assessment of soil colour. The system that is most often used is the Munsell notation. This has three components: *hue*, which indicates the major colour(s) present (i.e. YR is yellow-red, R is red, etc.); *value*, which is a measure of the degree of lightness or darkness of the colour; and *chroma*, which is a measure of colour intensity. A soil sample is then matched with the colour key. Thus a soil with a hue of 10YR, a value of 3 and a chroma of 4 would be classed as 10YR 3/4, which is yellowish brown. Other systems have been devised (e.g. Melville and Atkinson, 1985; Barron and Torrent, 1986) but the Munsell system is probably the most widely used. Three colour indices based on the Munsell notation have been proposed. In the Buntley–Westin index (Buntley and Westin, 1965), the hue is converted to a number (7.5YR = 4; 10YR = 3; 2.5Y = 2; 5Y = 1), which is multiplied by the chroma. Hurst (1977) devised an index by recalculating hue to a number (5R = 5; 7.5R = 7.5; 10R = 10; 2.5R = 12.5; 5YR = 15; 7.5YR = 17.5; 10YR = 20), which was then multiplied by the product of the fraction/chroma. The rubification index of Harden (1982) compares the colour of each horizon with that of the parent material, with a shift in the value of hue and chroma each being worth 10 points. Once the values of all three indices have been calculated for each horizon, they are multiplied by the horizon thickness.

Soil depth

Soil depth on flat surfaces usually reflects the length of time the soil has been forming. In similar environments, deeper soils tend to be older soils, as weathering at the bedrock/soil interface has changed more solid rock into soil components. However, because of depositional processes some alluvial soils are deep but very skeletal. Under stable conditions, and in situations where the rate of soil formation is enhanced, soils can be extremely deep. In some humid tropical areas, where weathering is very rapid, deep soils are found. In dry areas, where weathering and soil formation occur at a slow rate, soils tend to be shallow. Soil depth will also depend on the nature of the parent material and especially on slope steepness. Soils on limestones tend to be thin because solution removes much soil material, leaving only the insoluble residue. But it is slope angle that largely influences soil depth.

Soils are generally thinner on steep slopes because of the downslope movement of material. There is thus a relationship between the rate at which soil is formed and the rate at which it is removed downslope. Some of the soil that is removed will accumulate at the base of slopes, and this has led to the distinction between *accumulative* and *non-accumulative* soils. This distinction can also be seen in terms of a denudational balance, noted in Chapter 1. Therefore, the depth of soil may provide information about the soil's history.

CHEMICAL PROPERTIES

Chemical properties of soils are largely controlled by the reactions that occur between water in the soil and soil particle surfaces. The main reactions are *dissolution* and *precipitation* of salts and minerals, and *adsorption* and *desorption* on the surfaces of clay minerals, oxides and hydroxides of iron and aluminium, and humus. The composition of the soil solution is kept within a quite narrow range by *buffering*, which prevents excess leaching (see Chapter 3) and maintains a ready supply of soil nutrients. The main chemical properties are now examined.

Organic matter

Organic matter is mostly concentrated near the surface and varies in nature and amount. It varies from undecomposed plant and animal tissue to humus. Plant cells are made up of carbohydrates, proteins, fats and organic acids, lignins, waxes and resins. The bulk is composed of carbohydrates, but lignins are more stable than carbohydrates. Proteins decompose into amino acids. In soils, 5–30 per cent of the carbon exists as carbohydrates. The simplest and most easily broken down compounds are sugars and starches. More difficult to break down are hemi-cellulose and cellulose, the latter being the most abundant carbohydrate in plants. Proteins and amino acids (50–55 per cent carbon, 20–25 per cent oxygen, 15–20 per cent nitrogen, 6.5–7.5 per cent hydrogen) are important sources of nitrogen. It has been estimated that 20–50 per cent of all organic nitrogen in soils exists as amino acids.

Humus is a complex and resistant mixture of brown and dark brown amorphous and colloidal substances modified from the original tissues and synthesised by various soil organisms. It is the major component of soil organic matter. Much of the soil organic content adheres strongly to the mineral particles, especially clay, to form a **clay–humus complex**. Humification produces an organic colloid of high specific surface and high cation exchange capacity (see below). Its chemistry and the processes involved in its formation are extremely complex. Large amounts of carbon dioxide are produced during its formation, which increases soil acidity and enhances weathering. Organic acids also promote weathering. Carbon usually makes up over one-half of organic matter, and the carbon:nitrogen ratio is a good indication of the amount of decomposition of the original organic material. The ratio is high (>20) in plant tissue and low (<10) in humus. Soil organic matter is important in influencing the structure that develops in the upper horizons, and it takes part in many chemical reactions. It increases the water-holding capacity and the cation exchange capacity of soils.

Even after a century of study, the dynamics of soil organic matter (SOM) remain imperfectly understood. It is widely assumed that fresh plant detritus is converted gradually to humus and that this stabilisation involves a variety of physical, chemical, faunal and microbial processes. Most of the stabilisation mechanisms are not well understood, and most models of SOM dynamics divide SOM into several compartments based on assumed turnover rates; they define the dynamics in terms of transfers from one compartment to the next (e.g. Parton *et al.*, 1987, 1988; Jenkinson, 1990). A review has been provided by Sollins *et al.* (1996), who also propose a conceptual model based on stabilisation of fresh plant material into SOM and concomitant processes of destabilisation. The production of SOM is examined in greater detail in Chapter 3.

Humus can be divided into **mull** humus and **mor** humus. Under deciduous trees, a loose litter layer 2–5 cm deep develops, followed by a porous, dark brown soil that changes colour only gradually with depth. A deep organic A horizon, rich in soil animals, is characteristic of mull humus. Under coniferous trees, a thick layer, 2–20 cm deep, forms that is sharply differentiated from the mineral soil. The soil is capped by a thin band of blackish humus, usually compact, poorly drained and devoid of earthworms. This is mor humus. Mull humus also occurs under well-drained grassland and mor humus under heath vegetation. The type of humus will also depend on the nature of the parent material and its mineral status and on the calcium, nitrogen and phosphate content of the soil. Intermediate types, known as mor **moder** and mull moder, also occur.

Three organic layers can usually be identified. The *L layer* is undecomposed litter and is the organic matter deposited during the previous year. The *F layer* is the fermentation layer, partly decomposed with abundant fungal growth and faecal pellets. The *H layer* is the humified layer of completely altered plant residues and containing abundant faecal pellets.

Cation exchange

Soil colloids, a combination of clay and humus particles already referred to as the clay–humus complex, are also known as the *exchange complex* and whether inorganic or organic tend to possess net negative surface charges. As a result of this negative charge, colloidal particles behave like giant anions and are known as **micelles**. Cations associated with minerals are attracted to these charged surfaces (Figure 2.10). The strength of this attraction may lead to some cations being exchanged for others. In the majority of soils, the commonest cations are calcium, magnesium, potassium and ammonium. Cations with a high valency have a high energy of adsorption. The generally accepted sequence of preferred adsorption for base cations is $Ca^{2+} > Mg^{2+} > K^+ > Na^+$. There are increasing amounts of sodium in alkaline soils and hydrogen and aluminium in acid soils. Cation exchange and adsorption predominate in soils, but sometimes

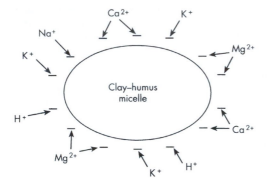

Figure 2.10 Clay–humus micelles and cation adsorption.

anion adsorption and exchange occur, especially in soils with limited humus content and in clays with a low specific area such as kaolinite.

The total negative charge on the surface is called the **cation exchange capacity** (CEC) and is usually expressed in milliequivalents (meq) per of 1 kg oven-dried material. Values range from 50 meq kg^{-1} for some lower soil horizons to 1,000 meq kg^{-1} for upper horizons rich in organic matter and/or certain clay minerals. Cation exchange capacity tends to increase with increasing organic content and with increasing content of some clay minerals, especially montmorillonite. Organic matter has an exchange capacity of about 3,000 meq kg^{-1}, kaolinite 3–15 meq kg^{-1}, illite 10–40 meq kg^{-1} and montmorillonite 80–150 meq kg^{-1}. Thus the CEC of some tropical soils is low because of large amounts of the clay minerals kaolinite and gibbsite. A high silt fraction may contribute substantially to the CEC. Oxides and hydroxides of iron and aluminium make little contribution to the CEC unless the pH exceeds 7.

The amounts of the various exchangeable cations filling the exchange capacity of a soil will depend on the inputs to and the outputs from the soil system. The main natural inputs are dissolution of minerals and atmospheric deposition. Artificial inputs occur by fertilisation and irrigation. The main loss to the system is by leaching. The cations with the least affinity for exchange sites are the ones most easily lost. Affinity depends on the charge and radius of the ion. Sodium is lost more easily than calcium, which is lost more easily than potassium. Nutrient cations are also cycled between soil and plant.

It was mentioned earlier that the soil solution is kept in a kind of balance by buffering. Buffering works in the following way. Precipitation water entering the soil by infiltration will dilute the soil solution. Whereas the concentration of hydrogen in rain water is generally greater than that in the soil solution, the concentrations of calcium and magnesium are lower. Thus hydrogen moves onto cation exchange sites, displacing calcium and magnesium into solution. This is the process of cation exchange. The soil system then tends to move towards an equilibrium between exchangeable cations and solution cations. If there is excess removal, exchange sites become decreasingly dominated by calcium and magnesium and the soil becomes acid. Buffering also occurs with respect to nutrient uptake by roots. Plants will take up those nutrients they require for growth. This will decrease the concentrations of those nutrients in the soil solution, and cation exchange will tend to restore the equilibrium. If potassium is being taken up by the plant, potassium will be released from exchange sites and will be replaced by other cations from solution. Roots take up a variety of cations and thus multi-ionic exchange occurs.

Related to cation exchange is **base saturation**, which is defined as the percentage of base **ions** (nonhydrogen) that make up the total exchangeable cations. This is calculated as follows:

$$\text{Base saturation (\%)} = \frac{(\text{Ca}^{2+} + \text{Mg}^{2+} + \text{K}^+ + \text{Na}^+)}{\text{CEC}} \times 100$$

The difference between CEC and total exchangeable base content provides a measure of exchangeable hydrogen content or exchangeable acidity:

$$\text{Exchangeable H}^+ = \text{CEC} - (\text{Ca}^{2+} + \text{Mg}^{2+} + \text{K}^+ + \text{Na}^+)$$

This means that there is a close relationship between base saturation and soil pH. The anion exchange capacity can also be determined and has been used for some tropical soils.

Soil acidity and alkalinity

It is the hydrogen ions in the soil that determine acidity. The acidity or alkalinity of a soil is measured by its **pH**, which is simply a measure of the accumulation of hydrogen ions. A pH of 7.0 is considered neutral. As pH is based on a negative logarithmic scale (pH = $-\log_{10}[H^+]$), low pH numbers indicate acidity and high pH numbers are indicative of alkalinity. The pH value of most soils ranges from 5 to 9. Very low values are found in soils of drained marshes and swamps that contain sulphur or pyrite. High values usually result from the presence of sodium carbonate.

Soil pH is usually measured in a standard suspension of 1:2.5 weight to volume, made up with distilled water, but sometimes a dilute solution of calcium chloride is used to minimise calcium release from the soil exchange complex. Values of pH measured in this way are generally lower than those measured with distilled water. Thus it is important to note the method of measurement.

Leached soils tend to be acid and are therefore associated with wetter climates. Soils are more alkaline in drier regions. The degree to which soils become acidic will depend on the inputs of acidity from vegetation, the microbial biomass and the atmosphere and also on the ability of primary minerals to resist the acidifying effects of leaching. The degree of alkalinity will depend on parent materials, vegetation and hydrology.

Hydrogen ions are derived from rainfall and from organic and inorganic acids produced within the soil. Root and microbial respiration produce carbon dioxide, which dissolves to form carbonic acid. Organic acids are a major source of hydrogen ions. Also, soil minerals are normally acid and release hydrogen during the weathering process. The exchange complex in acidic soils is dominated by hydrogen, aluminium and hydroxy-aluminium

ions, and base content is low. When the cation content increases, these ions are replaced and the pH increases. Alkalinity will depend on the strength of the base created. pH will not be constant within a soil but will vary within the soil body because of variations in carbon dioxide content, organic acid concentrations, the composition of exchangeable bases and the presence of roots, because they commonly contain adsorbed hydrogen ions.

Soil acidity may promote more acidity by helping the hydrolysis of Al^{3+} ions. This releases more H^+ ions into the soil suction by:

$$Al^{3+} + H_2 \rightarrow Al(OH)_2^- + H^+$$
(hydroxy-aluminium)

The hydroxy-aluminium may also suffer hydrolysis producing even more H^+ ions:

$$Al(OH)_2 + 2H_2O \rightarrow Al(OH)_3 + 2H^+$$
(stable gibbsite)

If soil pH falls below about 5.5, Al^{3+} ions begin to occupy exchange sites.

Soil salinity and sodicity

Soil salinity is primarily associated with arid and semi-arid regions. Under such conditions, there is insufficient water to leach away the soluble salts and they build up in the soil. Also, evaporation at the soil surface and transpiration by plants removes water but leaves the soluble salts in the soil. Irrigation with saline water also adds to the problem. The total salt content or salinity of water is determined by measuring its electrical conductivity. Analysis of saline soils also involves the measurement of sodium, calcium, magnesium and carbonate content.

Specific problems are created by high concentrations of sodium (**sodicity**). The amount of exchangeable sodium is expressed as an exchangeable sodium percentage (ESP), which is the amount of exchangeable sodium expressed as a percentage of the cation exchange capacity. Soils with ESP values above 10 or 15 are likely to exhibit special characteristics. Soil clays are liable to swell and disperse, leading to a breakdown of the soil structure. Heavy-textured soils become sticky and more plastic when wet and hard when dry. Hydraulic conductivity is decreased, leading to ponding on the soil surface, which makes it more difficult to wash salts out of the soil system. Air entry will also be restricted and anaerobic conditions may develop. The soil surface becomes sensitive to the mechanical effects of rain and to sprinkler irrigation, resulting in capping of the surface, which restricts infiltration. Sodic soils are usually classified as those soils with ESP values greater than 15 per cent. For a fuller treatment, see Chapter 8.

SUMMARY

Soils can be described in terms of their basic physical and chemical properties. Some of these properties can be assessed easily in the field, whereas others require careful sampling and then elaborate measurement under laboratory conditions. It is the soil properties that provide a clue as to the pathways of soil formation and the processes that have been involved in the development of that soil. These processes are examined in the next chapter. Some of the more significant properties, sometimes called diagnostic properties, have a significant role in many soil classification schemes. These schemes are examined in Chapter 5. Thus a clear definition of soil properties is vital to an accurate understanding of the nature of that soil.

ESSAY QUESTIONS

1 How can soil micromorphology aid an understanding of soil processes?

2 Examine the factors that may influence the moisture status of soils.

3 Why is organic matter an important constituent of soils?

FURTHER READING

Baize, D. (1993) *Soil Science Analysis: A Guide to Current Use*, Chichester: J. Wiley & Sons.

Brady, N.C. (1990) *The Nature and Properties of Soils*, New York: Macmillan.

Ellis, S. and Mellor, A. (1995) *Soils and Environment*, London: Routledge.

FitzPatrick, E.A. (1986) *An Introduction to Soil Science*, Harlow: Longman.

Krinsley, D.H. and Doornkamp, J. (1973) *Atlas of Quartz Sand Surface Textures*, Cambridge: Cambridge University Press.

Ringrose-Voase, A.J. and Humphreys, G.S. (eds) (1994) *Soil Micromorphology: Studies in Management and Genesis*, Amsterdam: Elsevier.

Rowell, D.L. (1994) *Soil Science: Methods and Applications*, Harlow: Longman.

Simpson, K. (1983) *Soil*, Harlow: Longman.

3

SOIL PROCESSES

INTRODUCTION

The production of the three-dimensional organisation of a well-developed soil requires the relatively consistent operation of a complex suite of physical, chemical and biological processes. The fact that soils all over the world contain recognisable characteristics implies that some processes are common to the development of all soils. What differs is the mix and relative strength of the processes.

It has been stressed many times that soils are complex systems reflecting interrelationships between processes and factors of the hydrosphere, atmosphere, biosphere and lithosphere. Soils evolve and develop over time. Sometimes this development involves gradual, continuous and sequential adjustments within the soil. This is what has been called progressive pedogenesis. However, many studies have shown that progressive pedogenesis is not the only pathway of pedogenesis. Interruptions to pedogenesis or even reversals frequently occur. This is termed regressive pedogenesis. These interruptions are created by changes to the controlling factors of soil formation (see Chapter 4). Changes of climate and vegetation can trigger a major change in the nature and operation of soil processes. More subtle changes, such as erosion and deposition, may produce similar effects.

Change occurs when a threshold is passed. A distinction can be made between extrinsic and intrinsic thresholds. An extrinsic threshold exists within the system but will not be crossed, and change will not occur without the influence of an external variable. Intrinsic thresholds involve changes without a change in an external variable. Many thresholds in pedogenesis are intrinsic, but extrinsic threshold changes can also be important.

The concept of pedological thresholds is comparatively well established, but still little is known about what types of pedological environment are conducive to promoting rapid soil development as opposed to retarding soil development. McDonald and Busacca (1990) have examined such issues in their analysis of soil formation on an aggrading surface. An extrinsic pedological threshold was created and crossed when pulses of aeolian sediment produced overlapping soil profiles. An intrinsic threshold was crossed when a rising zone of carbonate accumulation engulfed a former A horizon. Soils that form very rapidly if a pedological threshold is exceeded would lead to incorrect estimates of the age of that soil (see Chapter 4 for a fuller discussion of rates of soil development).

Many of these concepts have been incorporated into a model of progressive and regressive pedogenesis (Figure 3.1). Progressive pedogenesis

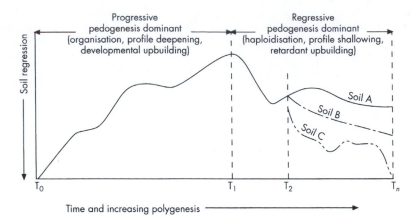

Figure 3.1 Hypothetical evolution of a soil using the concepts of progressive and regressive pedogenesis.
Source: After Johnson *et al.* (1990)

Episodic erosion eroded soils	Original soil	Episodic deposition composite and compound soils		
C	A C	A 2Ab 2Cb	A C 2Ab 2Cb	A C 2Ab 2Cb
Bw C	C	A 2Ab 2Bwb 2Cb	A C 2Ab 2Bwb 2Cb	A 2Ab 2Cb 3Ab 3Bwb 3Cb
	A Bwj or Bw C			
Bs Bw C	Bs Bw C	O or A Bs or Ej Bw C	A 2Bsb 2Bwb 2Cb	A 2Ab C or Bwj 2Cb or Bwjb 2Bsb 3Bsb 2Bwb 3Bwb 2Cd 3Cd
Bs Bw C	Bs Bw C	O or A E Bs Bw C	A 2Eb 2Bsb 2Bwb 2Cb	A 2Ab C or Bwj Bwjb 2Eb 3Eb 2Bsb 3Bsb 2Bwb 3Bwb 2Cd 3Cd

Increasing erosion ← → Increasing deposition

Increasing soil development →

↓ Increasing soil development

Figure 3.2 Development of eroded, composite and compound soil profile forms, Southern Alps, New Zealand.
Source: From Tonkin and Basher (1990)

includes those processes, factors and conditions that promote differentiated profiles leading to physical and chemical stability. Regressive pedogenesis includes those processes, factors and conditions that promote simplified profiles leading to physical and chemical instability and removals or retardant (non-assimilative) upbuilding. A good example of these effects has been provided by Tonkin and Basher (1990) for the Southern Alps of New Zealand. Soils on some slopes have been subjected to periodic phases of erosion and deposition, which have either retarded or enhanced soil formation. The various effects on the soil profiles are shown in Figure 3.2. Horizon nomenclature is examined in greater detail in the next chapter. Erosion leads to thin, relatively simple soils, whereas episodic deposition produces composite and compound soils.

The distinction between accumulating sites and degrading sites is sometimes blurred. Erosion will strip soils down to the C horizon. Excessive deposition will impose a new C horizon. On continuously aggrading surfaces, such as river floodplains, new C horizon material is added to the surface, the subsoil forms in a former A horizon, and the present A horizons of alluvial soils are formed in new superficial C material. Thus the genetic pathway is not C transformed to A or B directly but a new superficial C into a new A and a former A into a B. A former B horizon becomes a subsoil and is not to be confused with C horizons.

This brief introduction has shown that processes that operate within the soil can be grouped into those that result in additions (inputs) and losses (outputs), and those that lead to transformations and transfers (internal change or rearrangement) (Figure 3.3). The main additions are to the upper parts of soils from vegetation, water and dissolved elements from precipitation, particulate matter added by wind and water action, and by mass movement processes such as soil creep. Weathering of bedrock can also be considered as an addition, although most workers regard it as mostly a transformation process. Losses from the soil include surface soil erosion, throughflow and deep percolation of soluble components in water, and nutrient uptake by roots. Some of the nutrient uptake is returned to the soil in plant decomposition. There is also a gaseous and water loss by evapotranspiration. The main processes within each of these categories are now examined.

ADDITIONS

Additions are produced by *in situ* weathering of bedrock or parent material, by the inflow of material from upslope and by aeolian input. But production by weathering is best considered as a series of transformations. As was noted in Chapter 1, whether thin or thick soils develop will depend on the soil denudational balance. This has been embodied by Nikiforoff (1949) in the distinction between accumulative and non-accumulative soils.

It is recognised that aeolian additions are common in soils (e.g. Smith *et al.*, 1970; Syers *et al.*, 1969) and can be a major influence on soil development (Pye, 1987). Prominent horizons of carbonate accumulation occur in some desert soils formed in parent materials very low in calcium. Some of this calcium undoubtedly comes from atmospheric dust. In the Desert Soil Geomorphology Project of the US Soil Conservation Service in Dona Ana County, southern New Mexico, dust traps were used to assess the magnitude and nature of atmospheric input (Gile and Grossman, 1979). Clay content of the dust varied from 20 to 40 per cent, organic carbon ranged from 2.5 to 6.6 per cent and carbonate content from 1.3 to 5.7 per cent. It was concluded that calcium in dust was an important agent for the dissemination of carbonate over the landscape. Ionic calcium in the precipitation plus labile calcium in the dry dust would have been sufficient to form 2 kg m^{-2} of carbonate per thousand years, assuming that all of the calcium enters the soil and is deposited as carbonate (Gile *et al.*, 1981).

In the eastern Mojave Desert, California, Wells and McFadden (1987) have shown the way in which the accumulation of aeolian silt and clay has reduced soil permeability and lowered soil infiltration capacity. Aeolian deposition may also increase

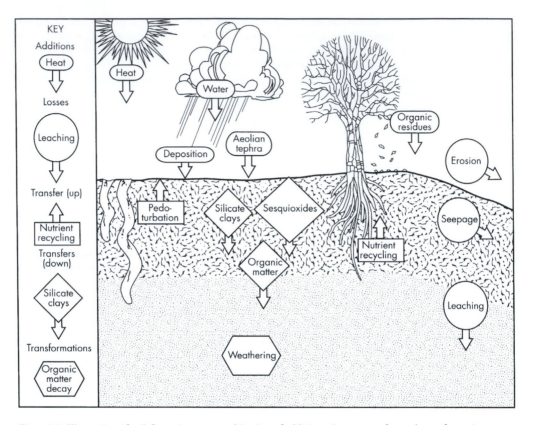

Figure 3.3 Illustration of soil formation as a combination of additions, losses, transfers and transformations.

hillslope instability by providing the fine matrix for debris flow activity. The presence of salts in the input will accelerate mechanical weathering (McFadden and Tinsley, 1985), as has been observed on basaltic lavas in the eastern Mojave Desert (Wells *et al.*, 1985).

Iceland provides a good environment within which to assess the influence of additions to the soil surface. Additions come from two main sources: volcanic ash (tephra) falls, and aeolian input of silt and fine sand as a result of wind erosion. Large tephra falls will blanket the entire soil surface and may be sufficiently thick to initiate a new phase of pedogenesis. Some tephra falls, such as that associated with the 1947 eruption of Hekla, have

been incorporated into the evolving soil and have had little immediate affect on pedogenesis. Peat bogs, excellent natural sinks, provide a record of the variation with time of windblown additions. A series of loss-on-ignition tests on samples from a peat bog in southern Iceland demonstrate the variability of this input (Figure 3.4). There is also some indication of increased inorganic content following major volcanic eruptions, suggesting that the stability of the landscape had been temporarily disrupted.

If additions are sufficiently large, stratification is created in soils, and it is important to be able to separate the sedimentological from the pedological characteristics. Several approaches to this problem

Scherpenseel and Kerpen (1967), using radioactively labelled clay, showed that much material was trapped in the upper soil layers and became permanently fixed. What little movement did occur appeared to do so in a type of pulse. Solution, however, appears to be a very effective transport agent, judging from the solute load of rivers. Removal of solutes from soil is influenced by rainfall frequency and intensity. Trudgill (1977) has identified four types of rainfall: low intensity/low frequency, low intensity/high frequency, high intensity/high frequency and high intensity/low frequency. Low-intensity/low-frequency rainfall will not be very effective, because, although solution will occur in the soil, little is removed from the soil. With low-intensity/high-frequency rainfall, water is supplied constantly enough for slow flow through the soil to occur as well as for solutes to be removed. High-intensity/high-frequency rainfall is the optimum for the removal of rapidly dissolving constituents but the opposite for the removal of slowly dissolving minerals. High-intensity/low-frequency rainfall will allow soil constituents to be dissolved as the residence time of the water is longer. This type of rainfall is probably more efficient than the second type because of a greater flushing effect. These considerations are not only important for determining the way the soil develops but are also significant in determining the fate of fertiliser and pesticide applications to soils (see Chapter 8).

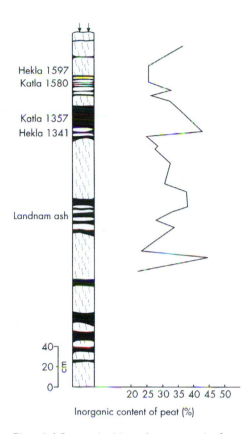

Figure 3.4 Loss-on-ignition values on samples from a peat bog at Ketilstadir, south Iceland.
Source: From Buckland *et al.* (1986)

have been suggested. The main principle employed is the disunity of the depth functions of soil properties, most especially particle size.

LOSSES

Material is lost from the soil in a variety of ways. Loss of soil material by erosion can be considerable and is examined in greater detail in Chapter 8. Water moves material out of the soil in suspension and by solution. Little is known about the amount of material passing through soil in suspension.

TRANSFORMATIONS

The main processes involved in transformations within soil are weathering, organic matter decomposition, the effects of waterlogging (including gleisation) and soil ripening.

Physical weathering

Physical weathering is the process that leads to brittle fracture of rocks and minerals. The main activating agents are unloading, thermal processes (thermoclasty) and a group of processes that involve the growth and expansion of material in pores and fissures, namely ice (gelifraction) and salts of various kinds (haloclasty). The nature and rate of physical weathering are influenced by some rock or material properties, especially porosity, water absorption, saturation coefficient, coefficient of volumetric expansion and thermal conductivity (Gerrard, 1988). Pore characteristics have always been thought to be important in determining resistance to salt and frost weathering. However, pore size and continuity are probably more important than total porosity. Rocks with high microporosity and high saturation coefficients are less resistant to physical weathering than rocks with lower values, even though total porosity might be similar. Other factors, such as tensile strength, permeability and mineralogy, will also be important, and a multivariate approach to physical weathering is desirable.

Unloading, or pressure release, is the creation of joints or fractures as erosion reduces the pressure on underlying rocks. Fractures (exfoliation sheets) created may be curved if the ground surface is curved, vertical or horizontal. Vertical joints or fissures created on vertical rock faces aid considerably the breakdown of rock and the retreat of the rock face. Unloading fractures are especially common in granites and massive sandstones but are also characteristic of many overconsolidated clays or mudstones when exposed at the surface by erosion. Some of the sheeting structures, especially those in igneous rocks, may be due to tectonic stresses.

The operation of insolation weathering (thermoclasty) is based on the low and extremely variable thermal conductivity of rocks. Low thermal conductivity prevents the inward passage of heat, causing the outer fringe of rock to expand preferentially against the colder core. This might lead to rock spalling. Minerals possess varying thermal conductivities, which will also generate stresses in the rock. The process appears intrinsically feasible, but early experiments (e.g. Blackwelder, 1933; Griggs, 1936) failed to replicate the process experimentally unless water was also present. There is no doubt that different rocks absorb heat differently. McGreevy (1985) conducted a series of experiments on chalk, granite, sandstone and basalt. The highest surface temperatures were recorded in basalt, reflecting its low albedo, low specific heat capacity and low thermal conductivity. The lowest temperatures were recorded for chalk, because albedo, specific heat capacity and thermal conductivity are high. Subsurface temperature differences increase in the order chalk, granite, sandstone, basalt. Sandstone, with a higher thermal conductivity, attained a higher subsurface temperature than basalt despite a lower surface temperature. The role of specific heat capacity is unclear. The maximum surface temperature ranking obtained of basalt > sandstone > chalk is the inverse of specific heat capacity values. The results show that different rock types could experience different temperature variations under similar exposure conditions.

Water freezing in a closed system increases in volume by about 9 per cent, and a maximum pressure of $2,115 \text{ kg cm}^{-2}$ is possible at $-22°C$. Pressures such as these seem capable of breaking rock by frost weathering, although it seems that freezing must take place from the crack or pore downwards, sealing off the system. Ice crystals also grow in size as water is withdrawn from smaller pores, leading to more pressure build-up. Frost weathering produces material of various sizes from individual grains to large frost-shattered blocks. More recently, frost action in rocks has been looked at more closely because of the possibility that hydration and salt crystal growth may also be involved. Dunn and Hudec (1966) found that the frost susceptibility of argillaceous dolomites and shales was inversely proportional to the quantity of ice formed. Later work (Dunn and Hudec, 1972) enabled carbonate rocks to be subdivided into frost-sensitive and sorption-sensitive types. Sorption-sensitive rocks fail as a result of stresses created by expansion of absorbed non-freezable water, a form of hydration. However, Fahey (1983) found that frost action on some New Zealand schists was three to four times more effective than hydration in producing material in the < 2 mm size range.

Salt weathering involves three groups of processes; the growth of salt crystals, the hydration of salts and the thermal expansion of salt crystals. Salt crystal growth is probably the most important form of salt weathering and can create considerable pressures. The most effective salts, in order, appear to be Na_2SO_4, $MgSO_4$, $CaCl_2$, Na_2CO_3, NaCl, $MgCl_2$ and $CaSO_4$. It is thought that salt crystal growth is a more effective process than insolation weathering, wetting and drying and frost shattering. Hydration forces created by anhydrous salts may be as high as those created by frost action. When temperatures are high during the day, salt crystals low in water of crystallisation may be formed that absorb water at night, forming higher hydrates. Winkler and Wilhelm (1970) calculated that high pressures can be created if salt hydration takes place over 12 hours in a sealed pore. It is also known that the coefficients of thermal expansion of many common salts such as sodium nitrate, sodium chloride and potassium chloride are much higher than most rocks, and expansion might be capable of causing splitting or granular distintegration (Cooke and Smalley, 1968).

An interaction between salt and frost weathering has been suggested to explain highly weathered rock in polar regions. Watts (1983) observed salt encrustations associated with woolsack-shaped boulders and salt concentrations in weathering pits. But these ideas are inconclusive. Goudie (1974) and Williams and Robinson (1981) have produced results which show that the presence of salts actually enhances frost weathering, whereas McGreevy (1982) demonstrated that some salts in solution can inhibit frost damage.

Biotic weathering is often treated separately, although the processes can be related to either physical or chemical processes. Biotic activity can cause or aid chemical processes, and physical effects can be significant. The drying out of colloids attached to mineral surfaces may pluck minute flakes from the crystal surface. The growth of roots in cracks and joints can produce significant stresses, although there are few measurements of the magnitudes involved. Root channels also provide avenues for water movement and the transfer of nutrients and dissolved chemicals.

Chemical weathering

The most important chemical processes are solution, carbonation, hydration, hydrolysis, oxidation and reduction, and organic complexing.

Solution is an essential process whereby minerals break down as they dissociate into their component ions. Solution is dependent on environmental factors such as temperature, redox potential (see below) and pH. The relationships between pH and the solubility of some common minerals are shown in Figure 3.5. An example of solution is the case of sodium chloride (NaCl):

$$NaCl + H_2O \Leftrightarrow Na^+ + OH^- + H^+ + Cl^-$$

The arrow, pointing in both directions, shows that the reaction can be reversed if the water is removed by evaporation. The sodium chloride will then be precipitated. The solution of carbon dioxide in water is another good example and leads to carbonation:

$$H_2O + CO_2 \Leftrightarrow H_2CO_3$$

The solubility of carbon dioxide depends on the concentration in the air (partial pressure) and the temperature of the solution. More carbon dioxide can be dissolved at higher partial pressures and for a constant temperature, but at higher temperatures, less can be dissolved. The carbonic acid (H_2CO_3) then reacts with minerals in the process known as **carbonation**. This is especially significant for limestone weathering with calcite ($CaCO_3$) and dolomite ($CaMg(CO_3)_2$). Thus:

$$CaCO_3 + H_2CO_3 \Leftrightarrow Ca^{2+} + 2HCO_3^-$$

and

$$CaMg(CO_3)_2 + 2H_2CO_3 \Leftrightarrow$$
$$Ca^{2+} + Mg^{2+} + 4HCO_3^-$$

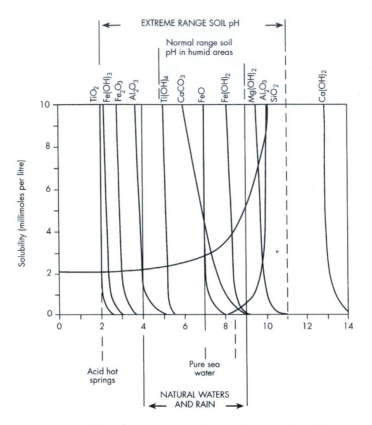

Figure 3.5 Solubility of some common soil minerals as related to pH.

the effect being to produce highly soluble calcium bicarbonate. In the weathering of feldspars, carbonation produces a clay residue of aluminium silicates and bicarbonate, which is usually lost rapidly in solution. Thus:

$$2KAlSi_3O_8 + 2H_2O + CO_2 \rightarrow$$
$$Al_2Si_2O_5(OH)_4 + K_2CO_3 + 4SiO_2$$

Hydration is the combination of a compound with water. The water is actually absorbed within the mineral lattice. Some minerals, such as gypsum ($\Leftrightarrow$ anhydrite) and haematite ($\Leftrightarrow$ goethite), are rapidly hydrated to form new minerals, often with a volume change. Thus:

$$CaSO_4 + 2H_2O \Leftrightarrow CaSO_4.2H_2O$$

and

$$Fe_2O_3 + H_2O \Leftrightarrow 2FeOOH$$

The process is reversible, and continual hydration and dehydration can cause physical weathering. As was seen in the previous chapter, some clay minerals, known as swelling clays, possess lattices that allow the easy ingress of water. The sodium-rich Mancos shales of Colorado increase in volume by up to 60 per cent under free swelling conditions. Wetting and drying cycles can produce distinctive soil features such as gilgai (see section on vertisols in

Chapter 5) and initiate soil creep. It has already been noted that hydration is one of the processes involved in salt weathering. Finally, hydration, by allowing water into mineral structures, enables other chemical processes to operate.

Hydrolysis occurs when metal cations in mineral structures are replaced by hydrogen ions in soil waters. Hydrogen possesses important characteristics that allow hydrolysis to occur and that make it such a fundamental process in weathering in general and in soils particularly. Hydrogen ions are small relative to other ions. They are almost the equivalent of subatomic particles, and such small particles always associate with other ions rather then occurring in isolation. Hydrogen is also unique in being able to form bridging bonds between negatively charged atoms such as oxygen. It is therefore able to migrate in chemical systems involving oxygen, and the way it does so enables metal cations to be released into solution.

The weathering of silicate minerals, especially feldspars, is largely by the process of hydrolysis, where the cations of the mineral are replaced by the H^+ ions of water and the OH^- ions combine with them to form solutions. Thus for albite (plagioclase feldspar), the reaction produces kaolinite and silica along with sodium ions:

$$2NaAlSi_3O_8 + 2H^+ + H_2O \rightarrow$$
(albite) $$Al_2Si_2O_5(OH)_4 + 4SiO_2 + 2Na^+$$
          (kaolinite)   (silica)

The concentration of H^+ ions is thus of fundamental importance. They are provided by the disassociation of a water molecule into hydrogen and hydroxyl ions, the dissolution of CO_2 with water to give carbonic acid, especially from the soil atmosphere, the production of acids by humus and biotic activity, the osmotic exchange of H^+ ions and nutrients, such as Ca^{2+}, Mg^{2+}, K^+ cations, plant roots, and the adsorption of cations at the surface and edges of clay minerals.

The products of hydrolysis may be moved out of the soil entirely, remain in solution, take part in cation exchange or form new clay minerals. Silica is slightly soluble at all pH values, but solubility increases rapidly at pH 9 due to the ionisation of H_4SiO_4 (Figure 3.6). Aluminium is not very soluble in the normal pH range of soils and usually forms clay minerals and hydroxides. However, aluminium does become soluble at low pH values, which is now being related to the action of acid rain and acidification (see Chapter 8).

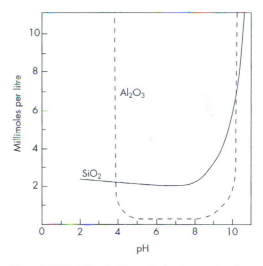

Figure 3.6 Solubility of silica and alumina as related to pH.

Hydrolysis, since it involves the movement of hydrogen ions, usually results in an increase in the pH of the solution. This is reflected in what is known as **abrasion pH**. If alkali-rich minerals are ground up in pure water, alkali cations are released and a rise in pH is observed. When equilibrium has been reached, the pH value is known as the abrasion pH. Feldspars have comparatively high abrasion pH values, for example orthoclase (pH 8), oligoclase (pH 9) and albite (pH 10). Micas range from pH 7 for muscovite to pH 8 for biotite. Carbonates yield pH 8 for calcite and pH 9 for dolomite, and quartz has a value of pH 7.

Oxidation/reduction involve the removal and addition, respectively, of **electrons**. The ability of a system to bring about an oxidation or reduction reaction is known as redox potential (E_h). Oxygen will accept electrons when present in the soil system, but if oxygen is unavailable, other elements act as electron respectors. The former situation is known as aerobic and the latter as anaerobic conditions. Anaerobic conditions are usually related to water-logged conditions in the soil. Thus oxidation occurs where there is easy access to the atmosphere and oxygenated waters. Well-drained soil conditions, higher temperatures and the destruction of organic matter all favour oxidation. The elements most commonly involved are iron, manganese, sulphur and titanium. Thus, both iron and manganese are divalent in their reduced state and increase to a valency of three or four, respectively, with the addition of oxygen and the loss of an electron. For example:

$$4Fe^{2+} + O_2 + 4H^+ \Leftrightarrow 4Fe^{3+} + 2H_2O$$

Iron and manganese are called ferric and manganic, respectively, in their oxidised state and ferrous and manganous in their reduced forms. Weathering occurs because in oxidation ions leave the lattice in order to maintain a balance. The lattice may then collapse or the vacancies be filled by other ions. Also, the ferrous states are often more soluble in the correct pH conditions.

Chelation, or organic complexing, is the formation of covalent bonds between metal atoms and organic molecules. This can make the metal soluble and is especially important for iron and aluminium solubility (see section on podzolisation in this chapter). This is why weathering can be very effective in organic-rich environments. It may even explain the increased weathering of lichen-covered surfaces, although physical processes are also likely to be involved in this example.

Clay mineral alteration

A variety of clay minerals can be produced in the soil, either by direct weathering from primary minerals or by transformation of one clay mineral into another. The ease of alteration depends on the degree to which the crystal structure has to change. Changes within and between 2:1 and 2:1:1 groups of clay minerals occur quite easily. Changes from 2:1 to 1:1 clay minerals are more difficult to achieve. It is not always clear what transformations will occur. This can be illustrated by the breakdown of feldspars (Gerrard, 1994). Feldspars have been noted, in individual studies, as weathering to illite, to kaolinite and illite, to smectite, to kaolinite and smectite, to smectite, kaolinite and several forms of halloysite, and to gibbsite. Much will depend on the micro-environment of weathering. Some clay minerals are produced directly from the weathering of feldspars as well as going through intermediate phases. In highly leached tropical soils, feldspars may be converted to gibbsite, either through an intermediate halloysite stage (Bates, 1962) or directly (Stephen, 1963). To complicate the situation, such a transformation may also occur in cool temperate (Wilson, 1969) or even alpine environments (Reynolds, 1971). Helgeson (1971) has postulated that feldspar hydrolysis involves the sequential sequence of intermediate products such as gibbsite, kaolinite and mica. Vermiculite and montmorillonite may be formed directly from feldspars or mica (Meilhac and Tardy, 1970).

Weathering indices

The progressive change of sound minerals to altered minerals has been used, in a series of weathering indices, to assess the degree of weathering. The indices are ratios of either chemical compounds (chemical weathering indices) or minerals (mineral weathering indices). Many of the chemical weathering indices use the amount of silica (SiO_2) or alumina (Al_2O_3), or both, in soils. Many alkaline earths are removed more easily than alumina. The more commonly used indices are shown in Box 3.1. The same principles underlie the use of mineral weathering indices. Information on the stability of minerals is used in constructing these indices. One of the most commonly used is the ratio of quartz to feldspar (Wrh); the higher the ratio the greater the

Box 3.1

THE MORE COMMONLY USED CHEMICAL WEATHERING INDICES

$$\frac{SiO_2}{Al_2O_3}$$ silica : alumina ratio

$$\frac{SiO_2}{Fe_2O_3}$$ silica : ferric oxide ratio

$$\frac{SiO_2}{Al_2O_3 + Fe_2O_3}$$ or $$\frac{SiO_2}{R_2O_3}$$ silica : sesquioxide ratio

$$\frac{K_2O + Na_2O}{Al_2O_3}$$ alkali : alumina ratio

$$\frac{CaO + MgO}{Al_2O_3}$$ alkali earth : alumina ratio

$$\frac{CaO}{MgO}$$ calcic : magnesia ratio

$$\frac{K_2O}{Na_2O}$$ potassic : sodic ratio

$$\frac{K_2O}{SiO_2}$$ potassic : silica ratio

amount of chemical weathering, since quartz is one of the minerals most resistant to weathering. Other indices employ heavy minerals. Olivine, amphiboles and pyroxenes are the least stable, while zircon and tourmaline are the most stable. Thus the ratio of zircon and tourmaline to the amphiboles and pyroxenes (Wrl) provides a good indication of the stage reached by weathering.

Organic matter decomposition

One of the most important transformations in soil is the decomposition of organic matter. Organic matter decomposition is the general term for a complex sequence of very detailed processes by which soil organisms use soil organic compounds as a food source. All living organisms can be classified on the basis of their main energy source into phototrophic (using solar radiation) and chemotrophic (using energy released from chemical oxidations). Each of these can be further subdivided on the basis of their principal carbon source. Autotrophs use inorganic carbon (CO_2), and *heterotrophs* use organic compounds such as carbohydrates. The four groups from these combinations and some examples are shown in Table 3.1.

Organisms mainly responsible for the decomposition of soil organic matter are chemoheterotrophs. These break down complex organic molecules to obtain energy and the simple nutrients they require to build their own tissues. Chemoheterotrophic soil organisms can be classified into groups according to their trophic roles (Ghilarov, 1970). **Phytophages** ingest plant material, **zoophages** are carnivorous and **saprophages** ingest dead organic matter.

Table 3.1 Classification of living organisms on the basis of energy and carbon nutrition

Category	Energy source	Carbon source	Example
Photoautotrophs	Solar radiation	CO_2	Higher plants; algae, blue-green algae, photosynthetic bacteria
Photoheterotrophs	Solar radiation	Organic compounds	Non-sulphur purple bacteria
Chemoautotrophs	Oxidation of organic compounds	CO_2	Specialised bacteria such as nitrogen- and sulphur-oxidising bacteria
Chemoheterotrophs	Oxidation of organic compounds	Organic compounds	All animals (vertebrates and invertebrates), most bacteria, most fungi

Source: After Ross (1989)

The rate of organic matter decomposition is controlled by the same factors that control microbial activity. Organic substrate quality and quantity are important because they influence pH and control soil moisture content. The most important environmental factor is soil moisture content, which influences soil aeration, temperature and leaching. It is suggested that the optimum temperature for organic matter decomposition and nitrogen mineralisation is 25–35°C. Thus, few soil micro-organisms operate optimally in temperate soils. Moisture affects microbial activity by limiting their movement in drought conditions, by controlling the transport of nutrients or toxins in the soil solution and to building up oxygen deficiency if waterlogging occurs. Soil moisture acts indirectly by breaking up the decomposable substrates by wetting and drying and freeze–thaw cycles. This exposes previously unavailable organic substrates for decomposition. Freezing is also lethal to the soil microbial biomass. This is probably the reason for increased rates of nitrogen mineralisation and phosphorus mineralisation after freezing and thawing.

The nature of the organic matter being decomposed is important. Molecular composition depends on species, the part of the plant or animal being decomposed and also age and season. The organic constituents of a range of plant materials are shown in Figure 3.7. As plants age, the proportion of cellulose, hemicellulose and lignin increases and the proportion of proteins and water-soluble constituents falls.

The 'quality' of plant material is an index of its palatability, its energy content and its nutrient value. It is determined by the amount of fibre or wood, the content of carbon compounds, which provide the energy source for decomposer organisms, and the content of nutrients such as nitrogen and phosphorus. Only specialist decomposers can break down lignin, lipids and chitin. Such decomposers are limited, which explains why these resistant compounds are particularly common in the surface organic horizons of soils. Thus van Cleve (1974) found a negative correlation between lignin content of tundra litter and rate of decomposition.

It has been suggested by Gray and Williams (1971) that the various stages in the decomposition process is like autogenic succession. Each wave of colonisation alters the substrate for the next wave. There is also a progressive depletion of the chemical energy source. Fleshy organic matter decays rapidly,

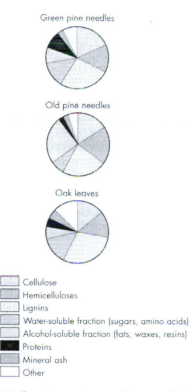

Green pine needles

Old pine needles

Oak leaves

☐ Cellulose
▨ Hemicelluloses
▢ Lignins
▨ Water-soluble fraction (sugars, amino acids)
▢ Alcohol-soluble fraction (fats, waxes, resins)
■ Proteins
▨ Mineral ash
☐ Other

Figure 3.7 Organic constituents of a range of plant materials.
Source: After Ross (1989)

losing water-soluble components, then carbohydrates such as cellulose and starch. The decay of woody tissue is much slower. Minderman (1968) has produced decomposition curves for a range of litter and soil organic constituents under Corsican pine (*Pinus nigra*), Scots pine (*Pinus sylvestris*) and sessile oak (*Quercus robur*). The composition by weight of the litter was sugars (15 per cent), hemicellulose (15 per cent), cellulose (20 per cent), lignin (40 per cent), waxes (5 per cent) and phenols (5 per cent). Almost all the sugars (99 per cent) and hemicellulose (90 per cent) were lost in the first year of decomposition. This was followed by cellulose (75 per cent), lignin (50 per cent), waxes (25 per cent) and phenols (10 per cent).

Ross (1989) has also described the breakdown of litter as occurring in a number of stages. In Stage Ia, many of the primary decomposers, such as beetle larvae, centipedes, millipedes and, in tropical soils, termites, comminute and physically break up litter detritus. In Stage Ib, a number of primary decomposers, such as saprophytic fungi and protozoa, which possess a wide range of extracellular enzymes, are able to dissolve outer protective tissue. The mechanisms of Stage I have made available a wide range of other organic substrates for further attack, in Stage II, by secondary decomposers such as most soil micro-organisms, saprophytic bacteria and faeces decomposers. In Stage III, the corpses and faeces from Stages I and II organisms provide a further diverse substrate for another hierarchy of decomposer organisms.

Much attention has been focused on carbon: nitrogen (C:N) ratios as indicators of rate of decomposition. However, Fogel and Cromack (1977) found that the lignin content of Douglas fir (*Pseudotsunga menziesii*) needle litter was more important than C:N ratio in determining the rate of decomposition. In an extensive study of the decomposition of heathland litter on deep peat in the Pennines, England, Heal *et al.* (1978) found that dry weight losses after one year (5–8 per cent for *Calluna* wood, 22 per cent for *Eriophorum* leaves, 25–45 per cent for *Rubus* and *Narthecium* leaves) paralleled decreasing fibre content, increasing total nitrogen (N), phosphorus (P) and potassium (K), and declining C:N, C:P and C:K ratios.

Substrates with high C:N ratios (of the order of 25–45), such as mor organic matter, are used by soil microbes as an energy source, carbon dioxide is evolved and nutrients, including nitrogen, are immobilised. This increases the nutrient content per unit volume and reduces the carbon/nutrient ratio. There appear to be critical C:N ratios of 20 for agricultural soils and 25 for organic soils, above which microbial immobilisation of nitrogen predominates and plants may suffer nitrogen deficiency.

The pH values of the litter and organic horizons are also important factors. Undecomposed organic horizons have low pH values because of carbon dioxide from respiring organisms and organic acid products of decomposition. Litter may also be acid, *Calluna* and *Sphagnum* have values of 4.4 and 2.8–4.4, respectively. Organic decomposition rates decline with low pH values. Collins *et al.* (1976) found that 80 per cent of aerobic bacteria in deep peat could not mobilise nutrients at pH < 5.5. Heal and French (1974) found that decomposition rates for litter of tundra species were lower on sites where the pH was below 4.5. A scale of grades for decomposition that may be used in the field has been provided by van Post (1922; see Box 3.2). It mainly uses the physical changes that accompany decomposition.

Waterlogging

Impeded drainage or very slow rates of soil water movement produce waterlogged conditions. This may occur because of clay texture, soil pans, compaction, smearing, the presence of a high local water

---

Box 3.2

VAN POST SCALE OF PEAT DECOMPOSITION

H1   Entirely undecomposed plant remains; squeezing in palm produces clear water.

H2   Practically undecomposed plant remains; squeezing in palm produces almost clear, yellowish brown water.

H3   Little decomposed peat; squeezing in palm produces dark-coloured water, but the peat, which is a very fibrous mass, does not protrude between the fingers.

H4   Poorly decomposed peat; squeezing in palm produces dark-coloured soil–water suspension. The plant remains are a little granulated.

H5   Somewhat decomposed peat; the structure of the plant remains is distinct to the naked eye, yet somewhat eroded. A little peat protrudes between the fingers on squeezing in palm, together with water in which large amounts of soil particles are suspended.

H6   Fairly well-decomposed peat; the structure of the plant remains is indistinct. On squeezing in palm, not more than one-third of the sample passes between the fingers. That part which remains in palm is granular and loose, and the structure of the plant remains is more distinct than in the wet and unsqueezed sample.

H7   Well-decomposed peat; the structure of the plant remains is still partially discernible. On squeezing in palm, about half of the sample passes between the fingers.

H8   Very well-decomposed peat; the structure of the plant remains is very indistinct. On squeezing in palm, about two-thirds of the sample passes between the fingers.

H9   Almost completely decomposed peat; the structure of the plant remains may only occasionally be recognised. On squeezing in palm, most of the sample passes between the fingers as a homogeneous soil–water mixture.

H10 Completely decomposed peat with no visible plant remains. On squeezing in palm, the entire sample passes between the fingers as a homogeneous mass.

table or surface ponding (Ross, 1989). The water-logging induces redox (reduction) reactions in soils and influences organic matter decomposition. The main effects have been listed by Ross (*ibid.*) as follows:

1 the production of gleying conditions in soils;
2 changes in pH;
3 the accumulations of organic matter due to slower anaerobic decomposition or fermentation processes;
4 the production of toxic by-products;
5 the influence of reducing conditions on retention, transformations and losses of native plant nutrients.

In general, there is a sequence of events when a well-aerated soil becomes waterlogged. Patrick (1978) has recognised four stages. In the first stage, NO_3^- reduction begins before complete removal of O_2. Reduction of Mn^{4+} to Mn^{3+} occurs during the reduction of O_2 and NO_3^-. The reduction of iron (Fe^{3+} to Fe^{2+}) is usually the second stage but will not occur if O_2 or NO_3^- is present in the system. Reduction of NO_3^-, Mn^{4+} and Fe^{3+} is undertaken by facultative anaerobic bacteria. The third stage is the reduction of SO_4^{2-} to S_2 (usually Fe_2S or H_2S) by true anaerobic bacteria and shows the complete absence of O_2, NO_2^- and NO_3^-. Finally, CH_4 is not produced until most of the SO_4^{2-} has been reduced to sulphide. Waterlogging has a marked effect on organic matter accumulation and decomposition. In the absence of oxygen, anaerobic fermentation occurs but is only about one-third as efficient as aerobic respiration. Soil micro-organisms utilise soil oxygen, which needs to be replaced, and if the rate of replenishment is slow, reducing conditions develop.

There are a number of reasons for the less rapid decay of carbon compounds under waterlogged conditions. These are a lack of electron acceptors for respiration processes, the production of end-products, such as hydrogen sulphide and ethylene (C_2H_4), which are toxic to soil micro-organisms, and the presence of higher concentrations of fatty acids, such as acetic acid, which inhibit microbial activity,

especially at low pH levels. Waterlogged soils are usually colder, which inhibits microbial activity, and possess lower pH values, which results in a less varied decomposer population.

The first products of anaerobic decomposition of the more readily decomposable organic compounds (e.g. carbohydrates) are methane (CH_4) and small amounts of H_2. Other gases, such as hydrogen suphide (H_2S), ethane (C_2H_6), ethylene, propane (C_3H_8) and propylene, can also be produced. Measurable quantities of ethylene have also been detected in well-aerated mineral soils (Dowdell *et al.*, 1972), and it seems that crumb aggregates can contain anaerobic centres even if the soil is not saturated.

Nutrient cycling is affected by waterlogging. Inorganic nitrogen in the form of $NH_4^+ - N$ occurs under waterlogged conditions. Nitrification, the production of nitrite ($NO_2^- - N$) and nitrate ($NO_3^- - N$) by the aerobic bacteria *Nitrosomonas* and *Nitrobacter*, is inhibited. Denitrifying bacteria thrive and the reduction of $NO_3^- - N$ is undertaken by *Thiobacillus denitrificans* to produce N_2O and N_2 as gaseous products. Nitrogen transformations are shown in Figure 3.8.

Gleisation

Intermittent or seasonal waterlogging and drying produces gley soils with characteristic mottles in one or more horizons (Figure 3.14). Mottles are usually orange-brown or grey-blue, indicating oxidised (Fe^{3+}) and reduced compounds (Fe^{2+}), respectively. Orange mottles are usually the result of the ferric orthohydroxides lepidocrocite and goethite. Blue-grey mottles have been ascribed to vivianite (ferrous phosphate) and ferrous sulphide. The production of these mottles is a biochemical process involving nitrogen-fixing bacteria such as *Clostridium*. As reduction and oxidation in gleys are microbially controlled, variations in moisture content and temperature effectively control the processes. The sequence of development of gley soils under temperate conditions has been itemised by Russell (1973):

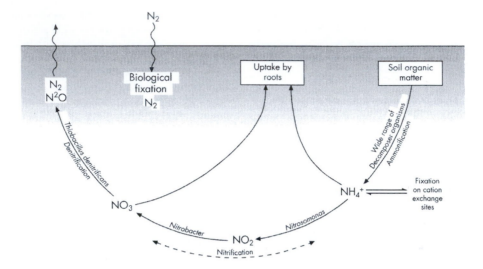

Figure 3.8 Soil nitrogen transformations within soil.
Source: After Ross (1989)

1 reducing conditions commence along old root channels in the autumn, when the soil becomes waterlogged but is still warm;
2 Fe^{2+} is produced in channels and diffuses into soil peds;
3 the following summer, this ferrous iron is oxidised to ferric hydroxide producing an orange colour inside the ped;
4 some Fe^{2+} produced during reducing conditions will be leached out of the soil;
5 Fe^{2+} may be moved into new soil horizons and deposited as ferric hydroxide when oxidising conditions arise.

Sulphidisation

Sulphur reduction is an important process in some soils (see Chapter 6) and is sometimes called **sulphidisation**. Estuarine and marine soils, especially in tropical regions, tend to be sulphur-rich because of sulphates in sea water. Sulphate-bearing water comes into contact with tidal marsh soils and submerged sediments (Figure 3.9). Bacterial reductions produce inorganic sulphides, usually hydrogen sulphide gas (H_2S), which reacts with reduced ferrous iron to form black ferrous sulphide (FeS), which may react with H_2S to form pyrite (FeS_2).

The sulphur content of tidal marsh soils increases with increasing organic matter content (Darmody *et al.*, 1977). This is not surprising, as organic matter is required in the sulphur reduction process. Pyrite formation in tidal swamps is favoured by tidal flushing (Fanning and Fanning, 1989). Tidal flushing removes HCO_3^- formed during sulphate reduction, producing a lower pH, which favours pyrite formation and also supplies small amounts of O_2 necessary for partial oxidation of sulphide to elemental sulphur, a necessary requirement for pyrite to form.

Soil ripening

Ripening is the name given to changes that take place in waterlogged or flooded soils when flooding is prevented and the soils dry out. Ripening can be physical, chemical or biological. Physical ripening

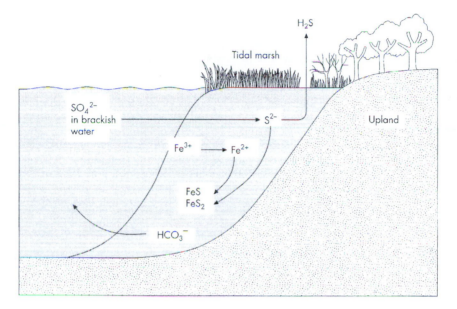

Figure 3.9 Sulphidisation of tidal marsh soils.

entails a considerable contraction of the soil mass as it dries out, creating numerous cracks. Air will then enter the soil and oxidation will take the place of reduction. This initiates chemical ripening with the oxidation of organic matter, iron and manganese. Biological ripening, sometimes called biological homogenisation, involves the mixing of the soil by the combined action of plants and animals. Flooded soils, such as those on floodplains and in estuaries, deltas and coastal swamps, are usually stratified, and this stratification is gradually destroyed. Changes are often most intense in the early stages of soil development on emerging salt marshes. An index illustrating the changes that take place during the soil-ripening process is provided by the *n* value of Pons and Zonneveld (1965). The formula is:

$$n = A - 0.2R/L + bH$$

where A is the total water content in grams per 100 g of dry soil, L and H are the percentages of clay and organic matter in the dry soil, R is the percentage of mineral particles other than clay (R = $100 - H - L$), and b is the ratio of the water-holding capacity of organic matter to that of clay. Values range from 0.3 to 0.4 for normal soils, whereas soft, freshly sedimented muds have values between 3.0 and 5.0. Mudflats exposed at low tide have values of about 2.0 and lowest salt marshes 1.2 to 2.0.

TRANSFERS

Transfers include all the processes that move material within the soil. This can be physical movement, such as mixing, or can be the movement of chemicals in solution. Transfers can be upwards as well as downwards in the soil.

Pedoturbation

Mixing of the soil, known collectively as pedoturbation, is important in most soils and, in some soils, can obliterate horizon development. Bioturbation is mixing caused by a variety of animals and organisms and is mostly confined to the topsoil. Humphreys (1994) found, from a survey of bioturbation near Sydney, Australia, that 95 per cent of activity was confined to the topsoil. Earthworms are one of the best-known organisms that mix and aerate the soil. A partial assessment of bioturbation from earthworm activity can be obtained by collecting and measuring surface castings. Lee (1985) has estimated that earthworms can produce over $20 \text{ kg m}^{-2} \text{yr}^{-1}$ of cast material. It is more difficult to assess subsoil activity, and most quantitative information is usually only in terms of casting. Subsoil activity can be determined where the rate of soil ingestion by worms is known (Satchell, 1958; Lavelle, 1978). Over half of the studies summarised by Paton et al. (1995) report rates of surface casting of $10–50 \text{ t ha}^{-1} \text{yr}^{-1}$ with higher rates generally being noted in wet/dry and temperate areas. Lower rates of activity occur in drier and montane areas. Some studies report rates of over $100 \text{ t ha}^{-1} \text{yr}^{-1}$, such as in Bangalore, India (Krishnamoorthy, 1985), northeast Thailand (Watanabe and Ruaysoongnern, 1984) and West Africa (Madge, 1969).

Ants, which are widely distributed in all climatic areas, generally live in large colonies and may burrow extensively, usually within 2 m of the surface. Some ant species produce quite large mounds. Humphreys and Mitchell (1983) have distinguished two types of mound. In type I mounds, material is simply deposited on the surface in the same way as earthworm casts. Such mounds are very susceptible to erosion. Type II mounds are larger, more compact, often cemented and resistant to erosion and therefore are present for longer periods of time. The funnel ant (*Aphaenogaster longiceps*), is an important bioturbator and produces type I mounds. Paton et al. (1995) report mounds varying from 5 to 25 cm in diameter and from 0.5 to 11 cm in height, with a mean volume of 240 cm^3 and an average of 190 g of mineral soil. The mounds were simply mineral soil piled as single grains or as fragile aggregates of up to four grains of fine sand linked by clay bridges. Rate of mound construction, essentially a summer process, was $841 \text{ g m}^{-2} \text{yr}^{-1}$.

The action of type II mound builders (*Camponotus intrepidus*) has been studied near Sydney, Australia (Cowan et al., 1985). These mounds are bigger and elliptical, covering individual areas of up to 1.2 m^2 and with average heights of 20 cm. Total volume, including nests below ground, ranged from 3,500 to $11,000 \text{ m}^3$. The soil was thoroughly mixed, with subsoil clay and topsoil and coarse organic debris being reworked into 5–8 mm clumps. Thus mounds are enriched in silt and clay compared with adjacent topsoil. Some ants even protect the mounds from erosion by laying down a cover of coarse material.

Quantitative estimates of ant activity vary considerably. Baxter and Hole (1967), studying prairie soil in southwest Wisconsin, found that mounds contained an average of $34 \text{ m}^3 \text{ ha}^{-1}$ and covered 1.7 per cent of the surface. An average occupancy rate of 12 years meant that the whole surface would have been occupied by mounds in about 700 years.

Some of the most impressive soil transfer is achieved by termites. Over 2,000 species of termite are known and are generally found in tropical and subtropical areas. Termites burrow and mould structures from soil to a greater extent than any other group of soil animals, and many build mounds. Pullen (1979) has noted that small mounds can occur at a rate of $5,000 \text{ ha}^{-1}$ in part of Zambia. Many of the large mounds are long-lasting, and there is not necessarily a large addition of soil each year. It has been estimated that most large mounds require no more than $100–590 \text{ g m}^{-2} \text{ yr}^{-1}$, less than the amount brought to the surface by earthworms and ants.

As they move and mix material, termites can produce new soil fabrics. Mineral fabrics occur in the outer walls of termite mounds and the infillings of former galleries, whereas organic-dominated fabrics occur in nursery areas and as gallery linings (Paton et al., 1995). Activity is most pronounced in drier environments, with upper values of mounding of about $1 \text{ t ha}^{-1} \text{yr}^{-1}$.

Other invertebrates that alter soil fabric and may produce mounds are beetles, wasps, bees, cicadas, crickets, grasshoppers, centipedes, scorpions, woodlice and crayfish (Hole, 1981). Isopods have also been noted as the source of much of the material subsequently removed by surface erosion processes (Yair and Rutin, 1981). A variety of other animals also move and mix soil. Some bird species create mounds and disturb soil by feeding. It was discovered that an average Australian lyre bird disturbed 6,300 g m^{-2} yr^{-1} during feeding, 4,500 g of which was mineral soil (Mitchell, 1985). It has also been suggested that 447, or 58 per cent, of mammalian genera could be involved in soil disturbance. Of these 447 genera, 31 per cent disturbed soil by surface feeding, 58 per cent by surface feeding and burrowing and 11 per cent by a subterranean lifestyle. Various human agricultural practices also mix and transfer soil, perhaps the most important being ploughing, which not only thoroughly mixes the soil but can create a hard plough pan at the depth of ploughing.

A number of non-biological pedoturbation processes can be described. Wetting and drying will redistribute soil and can lead to soil creep. Frost churning in soil brought about by repeated frost heaving and ice segregation within the seasonally frozen layer in periglacial areas is known as cryoturbation, which is responsible for a variety of patterned ground phenomena such as non-sorted circles and earth hummocks (Gerrard, 1992b). In such situations, buried peat and lenses of organic matter may be forced deep into the mineral material (James, 1970). This process is shown in Figure 3.10. Such processes decline in intensity in regions where thawing is deeper and winter snow cover greater. A variety of structures can be created within the soil, variously named festoons or involutions (Figure 3.11). Not all these features, however, can be related to repeated frost action within the active layer. Some may be due to gravity (loading) and related to permafrost degradation.

Infiltration

Infiltration is the process whereby water enters the soil. It should not be confused with the hydraulic conductivity of the soil, which is only one of several factors affecting the rate of infiltration. Infiltration capacity is the maximum flux of water across the soil surface. Infiltration rate is the volume flux across the

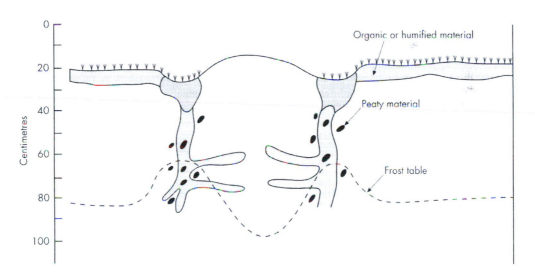

Figure 3.10 Illustration of pedoturbation created by frost heave and the creation of frost circles.
Source: After Ignatenko (1967)

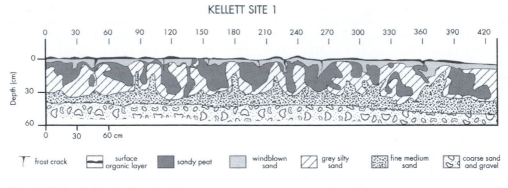

KELLETT SITE 1

| frost crack | surface organic layer | sandy peat | windblown sand | grey silty sand | fine medium sand | coarse sand and gravel |

Figure 3.11 Involutions in the active layer.
Source: After French (1996)

surface and will usually be less than the maximum possible value. A typical soil infiltration curve shows a period of rapid initial infiltration, which falls very quickly to a constant value. As soil moisture increases, saturation causes a reduction in the hydraulic gradient near the surface. Also, the presence of soil horizons of low permeability will shorten the time it takes a soil to reach saturation. Infiltration rate also decreases with time as a result of changes to the soil surface such as the reduction of pore size by clay mineral swelling or the washing in of fines.

If the capacity of the soil to receive water has not been exceeded, the amount infiltrated depends on precipitation rate and is said to be flux-controlled. Three modes of infiltration have been recognised; these are non-ponding, pre-ponding and ponded forms (Rubin, 1966). Pre-ponding occurs just before infiltration capacity is reached, and when rainfall rates exceed infiltration rates, ponded or profile-controlled infiltration occurs. Soil characteristics such as texture, structure, depth, nature and pro-portion of clay minerals influence the shape of the infiltration curve (Figure 3.12).

Percolation

The passage of water through soils is influenced by the size and arrangement of particles and voids. As

was noted in Chapter 2, the larger mineral particles form the skeleton of the soil, with the finer-textured materials acting as coatings on and bridges between the skeletons. This arrangement provides many pathways for water movement.

Water moves in two ways through soils. These are through large pores and by diffusion through soil aggregates. As water percolates through the soil it displaces water previously retained by the soil. At low rainfall intensities, much of the movement takes place through the fine pores in a process known as shunting. If the soil contains much water or rainfall intensities are high, water may move rapidly through large pores, known as macropores, in a 'by-passing' process. The length of time that water is in contact with soil will determine the chemical action it can undertake.

Water not only moves vertically but, unless the soil is on a level surface, will also move downslope within the soil in a process variously termed throughflow, interflow or lateral flow. Vertical flow tends to dominate in coarse-textured soils, but in fine-textured silts and clays, resistance to vertical flow may occur and throughflow is initiated. Whipkey (1969) measured water flow resulting from a simulated storm of 5.1 cm h^{-1} lasting two hours on a 16° slope. His results are shown in Figure 3.13. The small surface flow was caused by the low initial infiltration capacity of the dry surface soil.

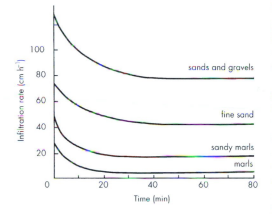

Figure 3.12 Typical infiltration curves for ponded infiltration on different materials.

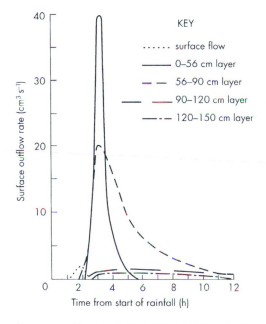

Figure 3.13 An example of discharge hydrographs for flow within different soil layers.
Source: After Whipkey (1965)

Infiltration rapidly increased with wetting. A less permeable zone was present in the soil at about 90 cm, and the lag before throughflow began represented the time taken for water to infiltrate to this layer.

An extreme type of macropore is known as a pipe and is responsible for the very rapid transmission of throughflow. A detailed investigation in the Wye catchment in the uplands of mid-Wales showed that some pipes were up to 0.6 m wide with a mean cross-sectional area of 67.5 cm² (Gilman and Newson, 1980). Soil and site properties such as drainage, depth to impermeable horizons, permeability and slope angle were used to derive an index known as the 'winter rain acceptance potential' (WRAP). This index enables the identification of soils most likely to allow throughflow and pipe flow. Soils most susceptible to piping were either those with peaty surface horizons and impermeable layers at shallow depth or those where surface horizons are loamy and the slope steep.

Pipes are generally produced by eluviation (see next section) and desiccation. Eluviation removes fine material from the soil matrix and possibly increases the size of pores by a mechanical effect, but it is only really effective in soils possessing initial weaknesses or low bulk densities. The literature on the creation of pipes by desiccation is large, but the precise manner in which the process operates is still uncertain. It appears to depend on the amount and type of clay minerals present. Parker (1964) has suggested that the enlargement of desiccation cracks was the principal mechanism of piping in clay soils, while eluviation was responsible for pipes in cohesionless materials.

Piping appears to be especially common in semi-arid regions, where many of the soils contain swelling clays such as **smectite**. Heede (1971) has suggested that a dispersive agent in the soil might be responsible for pipe development. He discovered that piping soils in Alkali Creek, Colorado, had a significantly higher exchangeable sodium percentage (ESP) than non-piping soils. Piping soils contained enough sodium to ensure dispersion when water moves through them and for swelling to occur. Pipes were initiated along cracks and, in soils with high ESP values, dispersion of soil into the cracks led to widening of the crack at its base when water ran through it.

Eluviation

Eluviation is the general term for the transfer of soil material in solution and involves two main processes, leaching and cheluviation. Leaching is the process whereby readily soluble soil components, both organic and inorganic, are removed in the percolating water. The processes involved in solution have already been examined. There is much evidence to suggest that **cheluviation** is a more important process. This is the process by which soluble organic complexes are responsible for the translocation of metal cations, especially iron and aluminium. Thus Grieve (1985) has shown that iron and aluminium are closely associated with dissolved organic matter in waters draining moorland soils.

Two main groups of organic material appear to be involved in cheluviation. The first are organic compounds such as polyphenols, which are washed directly from plant foliage and litter. The second group are condensed humic and fulvic acids and organic end-products of their decomposition. Malcolm and McCracken (1968), studying red oak, live oak and longleaf pine in North Carolina, estimated that about 20 kg ha^{-1} yr^{-1} of organic matter could be contributed by canopy wash alone. This amount of wash could mobilise up to 1.5 kg ha^{-1} yr^{-1} of soil iron and up to 0.7 kg ha^{-1} yr^{-1} of soil aluminium. Canopy drip from oaks was more effective in mobilising iron, while that from longleaf pine mobilised more aluminium. Fisher and Yan (1984) have also noted iron mobilisation by heathland species such as *Calluna* and *Erica*. It appears that polyphenols do the active complexing. Malcolm and McCracken (1968) also estimated that canopy drip from oak and pine produces about 1 kg ha^{-1} yr^{-1} of polyphenols. How the process operates is still not very clear. Schnitzer and Desjardins (1969) identified fulvic acid as the main organic

component (87 per cent) of the leachate from a humic podzol. This was substantiated by Dawson *et al.* (1978) in the soil solution passing through the A horizon of a podzol.

Lessivage is the movement of clays in collodial suspension without any change in chemical composition (Figure 3.14b). Lessivage tends to produce a group of soils known as acid brown earths, sol lessivés or luvisols (wash soils). Clay translocation occurs in many freely draining soils and is shown by the skins or layers of clay around soil particles or lining voids. Clay coatings have been examined in Chapter 2. Soil horizons characterised by clay accumulation are called argillic (see Chapter 5). Thus, argillic B horizons are used as evidence that clay translocation has occurred, although the form of the clay has to be established as there are a number of ways in which apparently clay-rich horizons can be formed.

Clay moves in suspension, therefore it must be first suspended and then maintained in a dispersed state. Wetting and drying soil may lead to disruption of the fabric and to dispersion of clay. Sodium ions in sufficient concentrations also disperse clay. This implies that the stability of natural soil colloids should be an important factor. Dixit *et al.* (1975) found that increased mobility occurred with an increase in the soil pH over the range 5.5–8.5 and with increased organic matter content. As pH rises, hydroxyl groups at clay crystal edges dissociate, increasing the colloid negative charge at its cation exchange capacity (CEC). Thus, since organic matter has a much higher pH-dependent CEC through the dissociation of carboxyl and phenol groups, we might expect organic colloids to be even more mobile than mineral colloids when solution pH is raised (Ross, 1989).

If clay is moving with percolating water, it will be deposited when water movement ceases. Water percolating in non-capillary voids will be stopped by capillary withdrawal into the soil fabric. The clay may then be deposited on the walls of non-capillary voids. It is also thought that if there is a zone of coarse texture underlying a fine-textured zone, water tends to remain in the fine capillaries above the zone

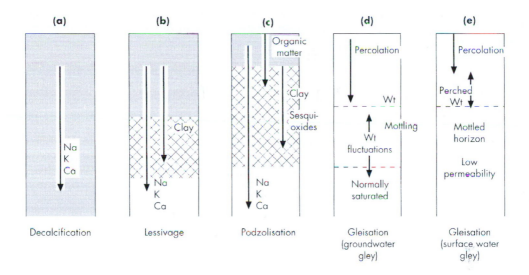

Figure 3.14 Main soil transfer processes. (a) Lessivage. (b) Decalcification. (c) Podzolisation. (d) Gleisation (ground-water gley). (e) Gleisation (surface water gley).

of contact. If this water is evaporated or withdrawn by roots, the clay may be left.

A number of specific leaching processes occur. Decalcification is the result of the leaching of calcium (Figure 3.14a). In some soils, the process may be balanced by the release of calcium from the weathering bedrock. The process whereby some or all of this calcium is redeposited is known as **calcification**. The term was originally used to explain the formation of chernozem soils (see Chapter 5), but many soils derived from calcium-rich parent materials exhibit signs of calcification. A lower soil accumulation of calcium carbonate is an indication that calcification has occurred.

Another distinctive process involving leaching is podzolisation, which occurs in freely drained soils on siliceous parent material, usually under a cool temperate climate where the vegetation provides a base-deficient litter that promotes the formation of raw acid organic horizons. Under such conditions, base cations are leached rapidly and are not replaced by other processes; consequently, clay minerals break

down and hydrous oxides of iron and aluminium are mobilised and removed from the upper horizon by the processes described under cheluviation (Figure 3.14c). This results in a bleached eluvial horizon. Some of this mobilised material accumulates lower down the profile to produce a diagnostic horizon enriched in iron, aluminium and organic matter.

Illuviation

The crucial question is how do clay minerals, iron and aluminium oxides and hydroxides and organic matter become accumulated? There are two basic groups of ideas. The first is a mechanical sieving effect, which will depend on the size of translocated material compared with the size of pore space in the lower part of the profile. The second, more controversial, group of ideas concerns chemical alterations of translocated collodial and chelated material, resulting in reduced solubility and precipitation. The illuviation of organometallic complexes may be the result of reduced acidity and biodegradation of organic matter, resulting in a higher metal:organic matter ratio and reduced solubility, thereby causing precipitation.

The surface charge or acidity of organic matter is a measure of the complexing activity and the solubility of its organometallic complexes (Buurman, 1985). Schnitzer and Skinner (1963, 1965) have demonstrated that a range of metal complexes can occur, ranging in metal:organic matter ratios from 1:1 to 6:1, and becoming increasingly insoluble as more metal is complexed. They also noted that, at higher pH values, higher amounts of metal are necessary to precipitate organic matter.

However, this idea of organometallic complexes being illuviated, especially in podzols (see Chapter 5) has been called into question by the discovery of imogolite-like allophanes (gel-like amorphous aluminosilicates) in podzolic B horizons (Farmer *et al.*, 1980; Farmer and Fraser, 1982). Wang *et al.* (1986) found that more imogolite or imogolite-like material occurred in Canadian podzols that were lower in organic matter than in soils that possessed higher organic matter contents. Thus, it has been argued that imogolite could not be deposited from humic complexes.

It is now thought that aluminium moves in the form of inorganic, proto-imogolite sols and not organic matter complexes. Other theories for organic matter deposition have been proposed (Anderson *et al.*, 1982). Acidic organic matter in soil drainage water may interact with previously deposited sesquioxide coatings to form precipitated organic matter complexes *in situ*. Also, colloidal organic matter may be precipitated preferentially at the top of the B horizon with deeper penetration of soluble fulvic acids. A summary of the two main theories of podzol development is shown in Boxes 3.3 and 3.4.

There is clearly considerable doubt concerning the accumulation of clay minerals and iron and aluminium oxides and hydroxides. This confusion is also present in discussions concerning the nature

---

Box 3.3

ORGANIC THEORY OF PODZOL FORMATION

Stage	Development process
1	Mobile organic substances are formed during the decomposition of surface litter and soil organic matter.
2	If there are sufficient Al and Fe cations at the top of the mineral soil profile, the mobile organic substances are immobilised and no downward migration occurs.
3	If insufficient amounts of Al and/or Fe are available to completely immobilise the mobile organic matter, these cations are complexed by the mobile organic matter and transported downwards.
4	Immobilisation of organo-mineral complexes may occur at depth due to: (a) supplementary fixation by cations; (b) desiccation; (c) on arrival at a level with different ionic concentration.

Source: After de Cornink (1980); Ross (1989)

Box 3.4

INORGANIC THEORY OF PODZOL FORMATION

Stage	Development process
1	At an early stage, before the development of A/E horizon, mobile Al, Fe and Ca are relatively abundant in the A horizon, and any fulvic acid freed by organic matter decomposition will be precipitated *in situ* as an insoluble salt.
2	Once an A/E horizon develops, fulvic acid becomes in excess of the A horizon capacity and moves down the profile. This fulvic acid can carry only limited amounts of complexed Al and Fe, but it will attack imogolite and proto-imogolite already deposited in the B horizon; this liberates silica and forms an insoluble Al fulvate *in situ*.
3	Thin iron pan horizons can provide effective barriers to the passage of fulvic acids – so in profiles with iron pans, imogolites persist up to the iron pan.
4	Finally, downward-migrating organic matter is sorbed on imogolite-like material of B_2 horizons.

Source: After Farmer (1982); Ross (1989)

---

and formation of duricrusts, referred to briefly in Chapter 2. It is only possible to outline briefly the main issues here, and readers are directed to books that have been written on particular types of duricrust (e.g. Goudie, 1973, on calcrete; McFarlane, 1976, and Tardy, 1993, on laterite; Valeton, 1972, on bauxite; Langford-Smith, 1978, on silcrete). There is also an excellent chapter in Ollier and Pain (1996). As was noted in Chapter 2, the generic name 'duricrusts' is given to the hard layers found in soils and sediments as a result of the transportation and precipitation of weathering solutions. Much discussion on duricrusts is bedevilled by the use of the term 'laterite', which is usually used to refer to a duricrust rich in oxides and hydroxides of iron, but the better term is 'ferricrete'. Some workers refer to a laterite profile, or lateritic weathering, or the process of **lateritisation**.

As noted, the character of duricrusts is determined by the concentration of specific material. However, as Ollier and Pain (*ibid.*) emphasise, it is necessary to distinguish between absolute and relative concentrations of material. Absolute accumulation implies that the abundance of a particular element increases by means of addition of the element from outside the system. Relative accumulation means that the element increases relatively by means of removal of other elements in the system. It seems that **silcrete** is always formed by absolute accumulation, because it is difficult to envisage aluminium being removed to give the relative accumulation of silica. **Calcrete** also appears to be formed by absolute accumulation. Ferricrete may be formed in either way.

There can be two types of ferricrete: pisolitic and vesicular. Pisolitic ferricrete consists of pisolitic or concretionary iron pellets, which may be dispersed, concentrated or cemented. Vesicular ferricrete has tubes, about 1 cm wide, partly filled with white kaolinite and red and brown oxidised material. This is probably the classic 'laterite' of southern India. The processes leading to the hardening of the crust have been described by Herbillon and Nahon (1988).

The definition of duricrust, noted in Chapter 2, emphasised the upward movement of iron from the base of the profile to the surface, where it is concentrated. Early suggestions included seasonal fluctuations in the water table. Evaporation of water from the top of a saturated column of iron-stained kaolinite was also suggested (Mann and Ollier, 1985). However, the main problem with these suggestions is created by the recognition that the weathering to produce the pallid zone did not release enough iron to account for the amount of extra iron in the crust. The addition of iron from windblown dust has been suggested (Du Bois and Jeffrey, 1955), but the most likely explanation is that the extra iron has been derived laterally. This is added recognition that the vertical movement of water and material in soils is not necessarily as important as was once thought.

SUMMARY

The processes that are influential in determining the nature of soils can be grouped into additions, losses, transformations and transfers. It is the transformations and transfers within the soil profile that are the most important and also the most complex. Examination of these processes demonstrates that there is still a great deal of uncertainty about the way these processes operate and the results of these processes. This is especially true of eluviation and illuviation. There are a number of external controls that determine the balance of additions, losses, transformations and transfers. These are examined in the next chapter.

ESSAY QUESTIONS

1 **Discuss the nature and significance of waterlogging in soils.**

2 **Examine the significance of pedoturbation in the development of soils.**

3 **Why is it that the podzolisation process is so little understood?**

FURTHER READING

Bland, W. and Rolls, D. (1998) *Weathering: An Introduction to the Scientific Principles*, London: Edward Arnold.

Courtney, F.M. and Trudgill, S.T. (1984) *The Soil: An Introduction to Soil Study*, London: Edward Arnold.

Foth, H.D. (1990) *Fundamentals of Soil Science*, New York: J. Wiley & Sons.

Goudie, A. and Viles, H. (1997) *Salt Weathering Hazards*, Chichester: J. Wiley & Sons.

Hillel, D. (1982) *Introduction to Soil Physics*, New York: Academic Press.

Jones, J.A.A. (1981) *The Nature of Soil Piping: A Review of Research*, Norwich: Geo Books.

Lee, K.E. (1985) *Earthworms: Their Ecology and Relationships with Soils and Land Use*, Sydney: Academic Press.

Lee, K.E. and Wood, T.G. (1971) *Termites and Soils*, London: Academic Press.

Martini, I.P. and Chesworth, W. (eds) (1992) *Weathering, Soils and Paleosols*, Amsterdam: Elsevier.

Paton, T.R., Humphreys, G.S. and Mitchell, P.B. (1995) *Soils: A New Global View*, London: UCL Press.

Ross, S. (1989) *Soil Processes: A Systematic Approach*, London: Routledge.

Singer, M.J. and Munns, D.N. (1991) *Soils: An Introduction*, New York: Macmillan.

Statham, I. (1977) *Earth Surface Sediment Transport*, Oxford: Oxford University Press.

Trudgill, S.T. (ed.) (1986) *Solute Processes*, Chichester: J. Wiley & Sons.

Trudgill, S.T. (1988) *Soils and Vegetation Systems*, Oxford: Clarendon Press.

4

CONTROLS ON SOIL FORMATION

INTRODUCTION

The operation of soil (pedogenic) processes is determined by a number of environmental factors operating over time to produce the soil at any one locality. These have come to be known as state factors, and when these are expressed as equations they become state factor equations. The equations should be considered as conceptual models or symbolic expressions, and it should not be thought that they can necessarily be solved in a mathematical sense, although some workers have attempted this with specific soil properties. The great Russian pedologist V.V. Dokuchaev appears to have been the first to propose such an equation when he identified the principal factors as climate, biota, parent material and landscape age or time. This can be expressed as:

$$S = f(cl, o, p)t^o$$

where S represents the entire soil or perhaps some specific soil property, cl is the influence of climate, o represents the range of organisms that contribute to soil formation, p is parent material and t^o represents relative age. It will be noted that relief or topography is not one of the factors that Dokuchaev

included in his equation. He did acknowledge relief as important, but mainly in the formation of what he called 'abnormal' soils. Dokuchaev, and many others in the influential Russian group of soil scientists, believed that broad regional soil patterns were determined by climate and biota – bioclimatic zones – producing zonal soils. This has undoubted similarities with the concept of climatic climax vegetation types (see *Fundamentals of Biogeography* by Richard Huggett). Such **zonal soils** were often called normal soils (Table 4.1). Soils largely affected by factors such as parent material, topography and drainage characteristics were called **intrazonal soils**. Soils that were poorly developed because of lack of time or because soil formation had been affected by significant erosion or deposition were termed **azonal soils**.

Shaw (1930) modified this Dokuchaev formulation as:

$$S = M(C + V)^t + D$$

which implies that soil (S) is formed from parent material M by the operation of climatic factors (C) and vegetation (V) over a time *t*, but that the process may be modified by erosion or deposition on the soil surface (D). This last point was very perceptive and,

Table 4.1 Zonal soils as classified by Dokuchaev

Zone	Soil
Boreal	Tundra (dark brown) soils
Taiga	Light grey podzolised soils
Forest–steppe	Grey and dark grey soils
Steppe	Chernozem
Desert–steppe	Chestnut and brown soils
Aerial or desert zone	Aerial soils, yellow soils, white soils
Subtropical and zone of tropical forests	Laterite or red soils

as was seen in Chapter 3, was incorporated into the idea of progressive and regressive soil development.

These early equations were later modified by adding relief to the soil-forming factors. The most significant development was the state factor equation of Jenny (1941):

$$S = f(\text{cl, o, r, p, } t, \ldots)$$

where S denotes any soil property, cl is the environmental climate, o represents animal organisms, r is relief, p is parent material and t is time since the start of soil formation. Jenny (1961) later modified the equation to make it more applicable to modern ideas concerning ecosystems. The revised equation is:

$$l, S, v, a = f(L_0, p_x, t)$$

where l is any property of the ecosystem, S denotes soil properties, v is vegetation and a is animal properties. L_0 represents the assemblage of properties at time zero, p_x is the flux of materials and t is the age of the system. The morphological characteristics of the system, such as its slope, exposure and topog-

raphy, is a subgroup of L_0, as are aspects of the mineral and organic matrix of the soil (in essence the morphological system). Climate is regarded as a subgroup of the flux potential, p_x.

Five broad groups of factors reflecting these five state factors are suggested (see Box 4.1). In this approach, t is assumed to be an independent factor, which is probably unrealistic. To remedy this, a series of dynamic models have been proposed. Wilde (1946) considered soil formation as a dynamic process with interdependent soil environmental factors interacting through time, i.e.:

$$S = f(G, E, B) \, dt$$

where G is parent material, E represents environmental factors interacting through time and B represents biological activity. This equation is still based on specific influencing factors. The equation was later modified by Stephens (1947), who divided

Box 4.1

STATE FACTOR EQUATIONS

The five groups of state factors, with the dominant factor placed first, can be arranged as follows:

$S = f(\text{cl, o, r, p, } t, \ldots)$ climofunction

$S = f(\text{o, cl, r, p, } t, \ldots)$ biofunction

$S = f(\text{r, cl, o, p, } t, \ldots)$ topofunction

$S = f(\text{p, cl, o, r, } t, \ldots)$ lithofunction

$S = f(\text{t, cl, o, r, p, } \ldots)$ chronofunction

A more convenient and perhaps realistic way of writing these is:

$S = f(\text{r})_{\text{cl, o, p, } t}$ topofunction

environmental influences into climate, c, relief, r, and water table, w, and changed the G and B factors into parent material, p, and organisms, o:

$$S = f(c, o, r, w, p) \, dt$$

which is essentially a dynamic version of the 'clorpt' equation.

A more generalised model was suggested by Simonson (1959):

$$S = f(A, R, T_1, T_2)$$

where A represents additions, R represents removals, and T_1 transformations and T_2 translocations to, within and from the profile. This is the general approach adopted in Chapter 3. Runge (1973) suggested an energy model, where water and organic matter were the organising and retarding vectors of pedogenesis. Thus:

$$S = f(W, O, t)$$

where W is soil energy related to leaching potential of water, O is organic matter production and nutrient cycling and t is time.

A rather more general approach, involving entire geospheres, has been proposed by Huggett (1995), resulting in a 'brash' equation. The geospheres involved are biosphere (life), b, toposphere (relief), r, the atmosphere, a, the pedosphere (soils), s, and the hydrosphere, h. Forces from the cosmosphere and lithosphere are excluded as being driving variables and not internal state variables. The spheres interact, so change in any sphere may be equated as a function of the state of all the spheres plus the effects of the driving variables, z. The pedosphere (soil) component becomes

$$ds/dt = f(b, r, a, s, h) + z$$

As Huggett (1995) stresses, this is essentially a version of the 'clorpt' equation with time included as a derivative. However, there is a fundamental shift in emphasis, although biosequences, climosequences

and so on could still be established, but the effects of individual terms on the right-hand side of the equation would require investigation with all other terms kept constant. But the soil system is one of a set, and singling out one state variable would probably produce erroneous results because the system acts as a whole. Thus 'great care should be taken when interpreting relationships between single soil and ecosystem properties and individual state factors' (*ibid.*: p. 36).

Paton *et al.* (1995) take a slightly different view. They argue that parent material, in the state factor model, was regarded as essentially influencing pre-pedological processes. They argue that near-surface processes of erosion and deposition exert a crucial control on the nature of the resulting soil material. Thus parent material is replaced by lithospheric material. This is similar to Runge's water-leaching potential. They also believe that climate is too vague to be of any real use and could be replaced with availability of water. This places the emphasis on the soil climate. The other factors of topography, biosphere and time do not require amendment. In terms of explaining soil variability over the time period of soil formation, lithospheric materials and topography would not vary to any great extent and therefore can be regarded as determinative factors in soil formation. Such determinative factors have a major difference in the way that they influence soil formation. Lithospheric materials are mainly involved in weathering processes and the production of saprolite. Topography exerts a control over near-surface processes and therefore the formation of topsoil. The availability of water and the nature of the biosphere have been subjected to greater variation over equivalent time periods. These factors drive the operation of the processes and therefore 'it is possible to think of these factors as operating an on/off pedogenic switch and to regard them as *initiating factors* in soil formation' (*ibid.*: pp. 109–10). Thus the formation of soil material results from the operation of weathering and near-surface processes within boundaries determined by the interacting factorial complex (Figure 4.1).

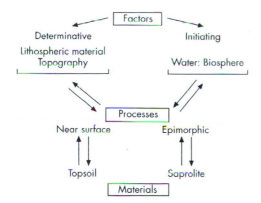

Figure 4.1 The pedological hierarchy of soil factors, processes and materials.
Source: After Paton *et al.* (1995)

Whichever 'model' of soil formation one adopts, there is reasonable agreement that factors of climate, biota, parent material and topography are important and that the length of time that processes have been operating influences the pathways of soil formation.

CLIMATE

Climate has a direct and indirect effect on soil formation. It acts directly by determining the moisture and temperature regimes within which soils develop. It acts indirectly through its role in controlling the processes of erosion, transport and deposition. Climate is important, because water is involved in most of the physical, chemical and biochemical processes that operate within soils. The amount of moisture influences weathering and leaching, and temperature influences the rate at which the chemical and biochemical processes operate. As an example, the rate of chemical weathering is thought to double with every 10°C increase in average temperature.

One of the major ways in which climate influences soil processes is through the soil water balance (Figure 4.2). Soil water storage (S), which is the quantity of water held in the soil water zone, is

increased by recharge during precipitation (P) and decreased be **evapotranspiration** (*E*). Any water surplus (R) is lost by downward percolation to the groundwater zone, by throughflow or by overland flow. The water balance can be written in the form of an equation as:

$$P = E \pm S + R$$

The difference between water use and water need is known as the soil water shortage (D), which will affect the growth of vegetation. A typical soil water budget for a humid mid-latitude climate is shown in Figure 4.3. Soil processes will be affected by the relative length of the periods of water surplus and deficit. Evaporation is usually expressed as potential evaporation (E_{pt}).

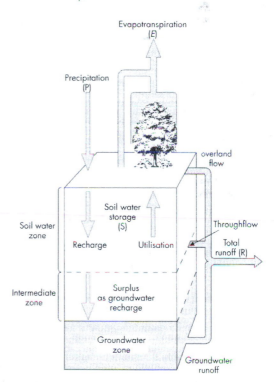

Figure 4.2 Idealised soil water balance.

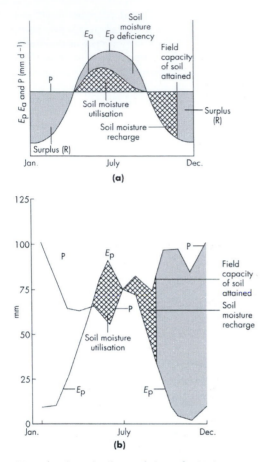

Figure 4.3 Annual soil water balance for (a) A hypothetical mid-latitude soil; (b) A soil in northern England.

If data on the water-holding capacity of a soil and on water balance are combined, it is possible to deduce water movement with depth in soils (Arkley, 1963), and to produce curves depicting water movement. A leaching index used by Arkley is the mean seasonal excess of $(P - E_{pt})$ for those months of the year in which $P > E_{pt}$. Soils in areas characterised by a high leaching index commonly possess properties associated with large amounts of water percolation, such as 1:1 clays. Regions with

low leaching indices and slight leaching tend to possess 2:1 clay minerals.

Climate also exerts an indirect influence through type and nature of vegetation and inorganic input into soils. It will be seen in the next section how biomass is related to broad climatically determined vegetation types. The growth rates of vegetation increase with temperature and humidity, but so does the rate of organic decay. The maximum rate of organic decay occurs between 25 and 35°C. Thus the input of organic matter into soils is determined by the balance between the rate at which material accumulates and the rate at which it is decomposed. In New Zealand soils, carbon content was found to increase from warm, dry areas to cool, wet regions as a result of decreasing decomposition relative to accumulation (Tate, 1992).

Climate acts indirectly by influencing the processes that add to or remove material from the soil. As a specific example, the nature and effect of the rain splash process depend on relations between rainfall characteristics and the nature of the soil and ground surface. Critical properties of rainfall are duration and intensity, and raindrop mass, size and terminal velocity (see Chapter 8). These variables determine the kinetic energy and momentum of rainfall. Thus most predictive models of rain splash action include rainfall intensity and duration parameters. Surface wash rates are influenced by a combination of climate and vegetation. Wash is very effective in dry savanna areas, but in areas of higher rainfall the increased vegetation cover gradually reduces wash to a minimum near the savanna–rainforest boundary. At higher rainfall amounts, the trend may be reversed because the controlling influence of vegetation cover is already at a maximum, whereas rainfall intensity may be increasing. There are three zones where runoff rates are high: the seasonal climatic zones of the mediterranean, monsoonal and semi-arid areas (Figure 4.4).

Wind speed is the main climatic factor determining the nature and rate of aeolian erosion and transport, but climate also acts indirectly by influencing ground surface conditions such as vegetation cover. Also, a dry soil is more easily erodible than a

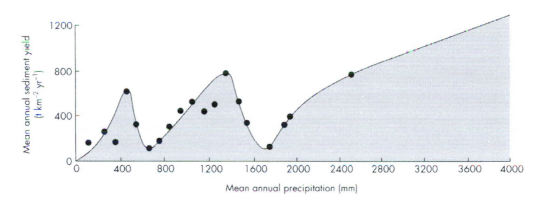

Figure 4.4 Sediment yield as related to precipitation and latitude (based on various sources).

wet one. Thus there is always a climatic factor in models to predict aeolian transport. The role of aeolian input will depend on the rate of that transport compared with the rate at which the material can be assimilated into the soil. Machette (1985) has produced an interesting model of these processes (Figure 4.5). Two general environments can be defined. Influx-limited regions are where

there is sufficient moisture to move material into and through the soil in either solid or dissolved forms. Thus most material deposited on the surface is moved into the soil. In moisture-limited regions, there is insufficient moisture, relative to influx rate, to move all material deposited at the surface into the soil.

The influence of climate on weathering can be seen by examining broad zonal divisions (Figure 4.6). The tropical forest zone is characterised by intense, deep weathering, with iron and aluminium oxides and hydroxides predominant and close to the surface. A variety of clay minerals are found at depth. Organic matter content is relatively low because of the high decomposition rates, which match the high accumulation rates. In deserts, there is a low organic input relative to the rate of decomposition; thus soils have low organic contents. Slight leaching produces predominantly 2:1 clay minerals, pedogenic carbonate at depth and gypsum at depth under extreme aridity. In the steppes, with increased precipitation and decreased evaporation, vegetation cover is denser, and A horizons rich in organic matter occur above moderately leached subsurface horizons. North of the steppes is the area known as taiga, with high soil leaching. Low temperatures result in low rates of organic matter decomposition. Where leaching moves iron and aluminium, an eluvial horizon can develop.

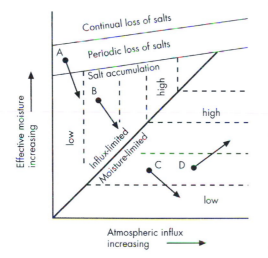

Figure 4.5 Pedogenic accumulation rates as a function of effective moisture and atmospheric influx.
Source: After Machette (1985)

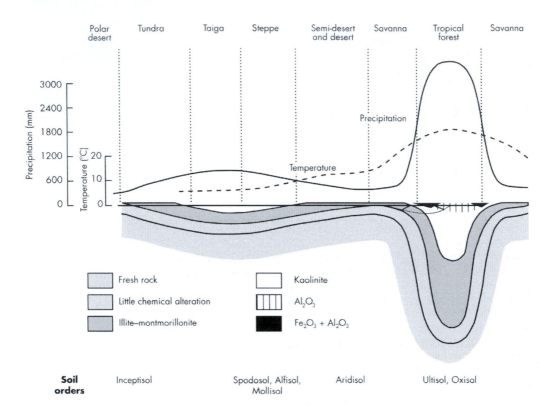

The periglacial environment (tundra, polar desert) is one of the most extreme for soil formation. Low temperatures and the presence of an impermeable, often perennially frozen, substratum are the major factors affecting the development of soils. It has even been suggested that the cold environment modifies soil-forming processes to such an extent that it may obliterate the effects of relief and time (Rieger, 1974). However, the periglacial climate can vary significantly. Tricart (1970) has distinguished three main climatic types: dry climates with severe winters and seasonal and deep freezing; humid climates with severe winters; and climates with a small annual temperature range with shallow and predominantly diurnal freezing. Permafrost is characteristic of the first type, is irregular in the second and absent in the third. Therefore, environmental conditions for soil formation will vary.

Most soils in periglacial environments are poorly drained and remain saturated or nearly saturated because permafrost prevents downward water percolation. Lateral movement is also reduced and can take place only in the thin upper horizons of soils. Soils consist of a surface organic mat, followed by a grey, greyish brown or bluish mineral layer, which is often mottled. The base is usually a grey frozen layer. Heaving and churning (pedoturbation) are normal characteristics of such soils, and the great variability in thaw and water availability, as well as variable microrelief, produce a complex soil pattern. The nature and distribution of organic matter are distinctive in such soils. Organic matter tends to accumulate on the surface and is not easily transferred into the soil profile, because of an absence of deep root systems. Organic matter also tends to accumulate immediately above the permafrost level, probably caused by the downward flow of colloids and dissolved humus. This humic alluvial layer may be unique to high arctic soils. The micromorphology of periglacial soils is also distinctive. Platy and lenticular structures are created by dehydration of the soil immediately below the freezing front as well as by compaction by frost-heave processes. Plasma coatings on grains are also a characteristic feature and have been ascribed to thixotropic behaviour of the soil mass during thaw. Skeletal grains may also become reoriented in the soil because of the downward penetration of the freezing front. As was noted in Chapter 2, many arctic soils also possess large bubble-like pores or vesicles. These can be as large as 2–3 mm in diameter and may be the result of thixotropic flow during thaw consolidation.

These rather extreme differences occur along what Tedrow (1977) has called a pedogenic gradient.

Arkley (1967) has attempted to show how some soil orders in the western USA can be grouped according to calculated actual evapotranspiration, leaching index and mean annual temperature (Figure 4.7). Ultisols and spodosols are associated with high leaching and aridisols with low leaching, and mollisols and alfisols occupy an intermediate position varying according to temperature (for descriptions of these soils, see Chapter 5).

Much has been written concerning the role of climate in chemical weathering, especially the possible relationship between climate and the specific type of clay mineral produced. Swelling clays such as montmorillonite are usually associated with arid and semi-arid climates, while clays such as kaolinite and gibbsite are thought to be associated with weathering under humid tropical conditions (Table 4.2). Although some studies seem to substantiate them, how reliable are these associations? Sri Lanka can be divided into wet and dry climatic zones, with a narrow intermediate band between. Gibbsite is a common clay mineral in the wet zone, whereas montmorillonite and kaolinite predominate in the drier zone and gibbsite is absent (Herath and Grimshaw, 1971). In the intermediate zone, minor amounts of gibbsite occur as well as montmorillonite. Except for the kaolinite in the drier zone, this

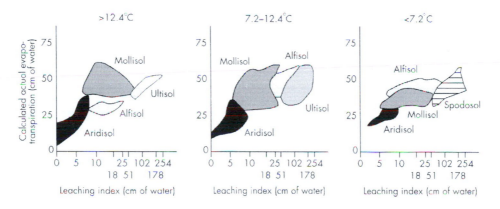

Figure 4.7 Relationship of soil orders to actual evapotranspiration and leaching index. Temperatures are mean annual values.
Source: Based on Arkley (1967)

Table 4.2 Clay minerals generally associated with different climatic zones

Climate	Clay minerals
Arid/semi-arid	Montmorillonite, chlorite, illite
Cool temperate humid	Montmorillonite, vermiculite, kaolinite
Warm temperate humid	Illite, vermiculite, kaolinite
Seasonally humid tropics	Kaolinite, gibbsite, illite, montmorillonite
Humid tropics	Kaolinite, gibbsite

generally fits the accepted 'model' for clay mineral formation.

While it is reasonable to assume that weathering rates under tropical conditions will be faster than under cooler conditions, there is no reason to assume that lateritic-type weathering cannot develop under other climates. This will make associating specific clay minerals with specific climatic characteristics difficult. The climate may only need to be wet enough and stable enough that developing profiles are not stripped. Taylor *et al.* (1992) report 1–3 m thick lateritic weathering profiles on early Tertiary basalt flows in the Monaro region of New South Wales. A zone of corestones grades up into a mottled clay zone, which gives way to a less structured zone with concentric red and yellow bands, suggesting former spheroidal weathering shells. There is an upward increase in the amount of gibbsite. Goethite is a dominant oxyhydroxide, with minor amounts of haematite and anatase. The transition to ferruginous crust is very abrupt. Studies of fossil pollen, spores and wood have shown that the climates were cool, wet and thermally seasonal, and the vegetation was cool to cool temperate rainforest similar to that currently in Tasmania, New Zealand and the highlands of New Guinea. It is suggested that approximately half a million years was needed for the development of the profiles. In this instance, the long period of development has overcome the deficiencies of climate. Also, as laterites formed in higher latitudes mostly occur on basalt, there appears to be parent material control as well. Iron-rich rock requires less restrictive conditions for such profiles to develop (Barron, 1986). Other studies have shown that kaolinite occurs in soils and weathered material in arid and semi-arid areas, and the temptation is to assume that this indicates a climatic change. But this may be a circular argument.

It is important to establish that similar parent materials are involved. The biotic component also influences clay transformation within soils. Spyridakis *et al.* (1967) have shown how the rhizo-spheric activity of coniferous and deciduous trees affected the transformation of biotite to kaolinite. In New Zealand, McIntosh (1980) found halloysite in more profiles under *radiata* pine than under native *manuka* scrub. It is clear that it is the specific leaching conditions at the mineral surface that determine the weathering pathway. Such leaching conditions will be governed by broad climatic characteristics such as temperature and moisture availability but are more likely to be determined by specific site characteristics. A number of studies have shown that clay mineralogy changes systematically along some soil catenas as a result of changes in water movement and intensity of leaching. The micro-environment of mineral weathering is the key to an understanding of clay mineral formation. However, it is possible to make a number of general conclusions (Gerrard, 1994). Montmorillonite may be the best climatic indicator, as it is rarely found in quantity in areas that have not experienced arid or semi-arid climates. Kaolinite would be expected to be the dominant clay mineral in deeply weathered zones in humid tropical areas, but it can also be produced under a variety of conditions. Halloysite may be a better indication of weathering under humid tropical conditions. Gibbsite, in appreciable amounts, probably suggests lateritic-type weathering, but small amounts are possible under a variety of conditions. If mafic minerals are being weathered, chlorite and smectite might be expected, especially under arid and semi-arid conditions. Illite is a poor indicator of environmental conditions. Also, a single clay mineral type might not be diagnostic, and the mix of minerals may be more significant. However, it is clear that great care is required in inferring climatic conditions from clay mineralogy. Confirmation will be required from independent sources.

This very brief review has shown that climate is a major factor in influencing soil formation but that detailed consistent relationships are difficult to establish. Thus 'when one compares some soil properties in quite different climates, pedologic differences that one might predict from the climatic data are not always apparent' (Birkeland, 1984: p. 275). Birkeland (1990) has summarised a variety of studies to examine clay and Fe accumulation with time under a variety of environments (Figure 4.8a). For clay accumulation, Group A includes soils from a diverse group of areas such as coastal California, the Great Valley of California, Washington state, Missouri and Barbados. Group B plots are characteristic of the Sierra Nevada and the arid and semi-arid areas of the western USA, Montana and the Rocky Mountains. Cp represents coastal plain soils of the eastern USA. The Fed accumulation data (Figure 4.8b) fall into Group A (California and some arid and semi-arid soils) and Group B (Great Valley, Rocky Mountains and other arid zones), with Cp again representing the eastern US coastal plain. As these trends are relatively consistent over diverse geographical areas, the role of climate is not clear. Similar confusion occurs when profile development index (PDI) values are plotted against time for different climatic environments. In a study that included climates ranging from arid to humid, most of the data cluster together and no pattern emerges (Harden and Taylor, 1983).

This uncertainty means that there is a great danger in trying to reconstruct climatic conditions from the properties of fossil soils. Colour is a good example to consider. Soils possessing deep-reddish horizons are often assumed to have formed in warm, humid climates, but the colour may simply reflect the length of time during which the processes creating the colour have been active. Greater

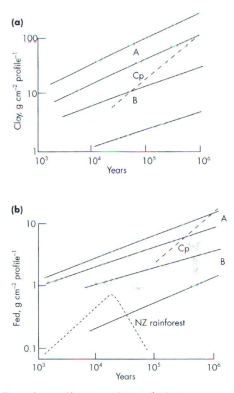

Figure 4.8 (a) Clay accumulation. (b) Iron accumulation with time in various environments. *Source*: Based on a summary by Birkeland (1990)

significance should be attached to those processes in which a threshold level of activation energy must be achieved for a reaction to take place. An example of this dilemma is the attempt by Rutter *et al.* (1978) to establish the difference in climate between two interglacials in the central Yukon by examining the type of clay minerals and their depths in two buried soils. Mixed montmorillonite–kaolinite clay minerals were found at depths of 190 cm in one soil, whereas vermiculite–chlorite intergrades were found at depths up to 93 cm in another. Using a formula devised by Birkeland (1974) to establish the amount of water needed to reach the two depths, the results suggested that the climate in which the first soil developed was probably considerably more humid than for the second. However, differences in depth of water penetration may simply be a function of time. There is an additional problem with this interpretation. As noted earlier, montmorillonite is usually associated with drier climates. Until specific relationships are established between specific climatic parameters and soil properties, such a process is bound to be speculative. Whether this will ever be achieved, considering the complexity of factors and processes involved, is doubtful.

BIOTA

The biotic factor is difficult to assess because of the dependence of both biota and soil on climate and the two-way interaction between soils and vegetation. Biota includes both plants and animals, but the emphasis is usually on vegetation. However, a range of organisms, from bacteria to large mammals, can affect soils, and their influence can be significant. FitzPatrick (1986) has argued that nearly every organism that lives on the surface of the Earth or in the soil affects the development of soils in one way or another. Organisms add, transform and transfer organic matter. They also transfer and transform mineral material and may facilitate the operation of other soil processes. This was seen with respect to bioturbation in Chapter 3. Many chernozems possess extensive networks of crotovinas produced by blind mole rats. Muskrat and pack rat middens, in different environments, can have an effect on soil properties. The same can be said of pocket gophers (Litaaor *et al.*, 1996). Concentrations of total C, N, exchangeable Ca and K were found to be significantly lower in gopher mound soils, whereas available NO_2 and nitrogen fluxes were significantly higher. Silt and clay contents were significantly higher in surface horizons compared with subsurface horizons of inter-mound areas. Particle size distributions were more homogeneous in the mounds. These differences were explained by the mixing activity of the gophers.

The influence of soil organisms depends on their life habit. Decomposers live in the surface organic horizon and ingest only organic matter. Burrowers live in the subsurface mineral soil and ingest both mineral and organic matter. Grazers live on the soil surface and eat vegetation, and predators live either on the surface or in the soil and prey on other organisms. Decomposers, burrowers and grazers transform organic matter directly. Predators return organic matter to the soil by excretion. Organisms also affect gaseous inputs, especially carbon dioxide, leading to soil acidification. Organic acids can cause hydrolysis, and organometallic complexing can take normally stable iron and aluminium compounds into solution, as was described in the previous chapter.

There are physical as well as chemical effects. Pedoturbation by soil organisms was examined in Chapter 3. Mando *et al.* (1996) have demonstrated that cumulative infiltration amounts were all greater on termite plots in northern Burkina Faso.

Most studies emphasise the importance of vegetation cover and type on soil formation and, at a very specific level, sites can be found where vegetation appears to be the most important factor. But there is always the possibility that microclimatological effects created by the vegetation cover may be more important. Vegetation shelters the surface from wind (evapotranspiration) and precipitation, and leaf drip and stemflow can be important.

An important effect of vegetation is the production of plant litter. Organic matter can also be produced by plant roots. Litter production is related to broad bioclimatic regions, with the greatest amounts being produced in tropical forests and least in tundra areas (Table 4.3). Litter can also be transported by wind and water and sometimes by animals, but it is always greatest near its source. The transformation of organic matter is also determined by the type of vegetation. Vegetation with a low content of nutrients and high levels of phenols, waxes and lignins will be less easy to decompose (Minderman, 1968). Needle-leaf trees produce mor humus with high C:N ratios, whereas broadleaf trees produce moder or mull humus with lower C:N ratios (Klemmenson, 1987; Bernier *et al.*, 1993).

Table 4.3 Input rates of litter under differing vegetation cover

Vegetation community	Litter input (t ha^{-1} yr^{-1})
Tundra	1.5
Boreal forest	7.5
Temperate deciduous forest	11.15
Temperate grassland	7.5
Savanna	9.5
Tropical forest	30.0

Source: After Swift *et al.* (1979)

Experiments in the San Dimas experimental forest, southern California, investigated the 2:1 phyllosilicate mineralogy of lysimeter soils under a 41-year-old monoculture of scrub oak (*Quercus dumosa*) and Coulter pine (*Pinus coulteri*). Mica content increased relative to vermiculite in both soils, but the increase was far greater under oak. Non-exchangeable potassium increased by 23 per cent in the clay fraction of the oak A horizon and 5 per cent under pine relative to unaltered parent material (Tice *et al.*, 1996). More potassium may have been fixed by vermiculite in the oak A horizon due to greater potassium concentration in the oak litter pool, earthworm-mediated mineralisation of potassium from organic matter under oak and the presence of fewer roots at the surface under oak, therefore less plant removal of potassium from the A horizon. These results indicate the significance of biocycling in soils.

In order to be able to assess the general role of vegetation, all factors other than vegetation need to be kept constant. This is possible only if we can assume constant regional climates. Many workers stress the major differences between soils developed under grasslands and those under forest. Studies conducted on forest and grassland soils in mid-North America have shown that all soils appear to possess an equally high content of organic matter at the surface, but the distribution with depth varies with vegetation. A horizons under trees are relatively thin, and organic matter content decreases rapidly with depth. Grassland A horizons tend to be thick, and the organic matter content remains high for some depth. This is partly a function of the way in which organic matter enters the soil. In forests, the main input is by leaf litter fall to the surface, while in grassland situations, organic matter enters the soil not only from litter fall but also from root decay at depth.

These differences in the nature and amount of organic matter input appear to determine soil properties. Percentage base saturation is higher in grassland soils, partly as a result of less leaching compared with forest soils. There are lower evapo-transpiration rates under a forest canopy, therefore there are more chelating agents and more acid leaching waters. There is also greater annual biomass production and cycling of cations in grassland soils, producing higher pH values than in forest soils.

However, oddities can occur. In parts of Wisconsin, the pH of surface soil under hardwood forest is maintained at a value of 6.8 because the tree roots are able to penetrate the underlying calcareous material and extract calcium, which is returned to the surface in litter. In contrast, the pH of surface soil under tall grass vegetation is 5.5 because the calcareous material is below the zone of the grass roots.

There are other significant differences between soils under grassland and forests in temperate environments. The ratio of Fe_2O_3 in the A and/or the E horizon relative to that in the B horizon is lower in forest than in grassland soils. This may also be a function of the chelating ability of the organic compounds. Grassland soils appear to possess higher clay contents in their surface layers than forest soils. If the soils are developed on similar parent materials, this suggests that clay particles can be translocated to greater depths in forest soils. Grassland soils have more carbon, with about twice as much nitrogen and half the carbon:nitrogen ratios of forest soils (Ugolini and Schlichte, 1973). The types of aluminium and iron also differ.

These differences pose an interesting question (Birkeland, 1984). If the forest–grassland boundary were to shift, what would be the effect on soils? Initially, one would expect transitional soils to develop, but the nature of these soils will depend on whether the transition is from grassland to forest or from forest to grassland. Organic matter and base saturation values might be the same in either case. If the change was from grassland to forest, iron and clay particles would probably start to move down through the soil. If the transition was from forest to grassland, the iron and clay 'fingerprint' might remain and could be used to infer former conditions. The E horizon of the former forest soils might persist if grass roots were shallow. Other soil properties might indicate that such a change had occurred. Some of the silica absorbed by plants is precipitated in plant cells to form opaline substances known as opal phytoliths. Grasses are the highest producers and produce phytoliths that can be distinguished from most other plants. Modern forest and grassland soils differ in their types of phytolith, and these differences can be used to infer gross vegetational changes (Jones and Beavers, 1964).

The effect of trees, in an individual sense, can be seen by noting the way in which soil properties change with distance from that tree (e.g. Zinke, 1962). Nitrogen content, pH, exchangeable bases and CEC values are all low near tree stems and increase to a maximum some distance from the crown before they start to decline. This seems to reflect the influence of bark and leaf fall near the trunk and the role of stemflow. Bark litter is predominant near the trunk and is more acid with a lower cation and nitrogen content. Furthermore, soil near the trunk can receive five times the water of soil in the open. Stemflow water also seems to have a higher content of chemical elements. The relative influence of stemflow and litter type varies with tree type. Trees with smooth bark have high stemflow, trees with rough bark generate low stemflow. Thus soil variation may be as much a result of amounts and nature of stemflow as the type and amounts of litter fall (Gersper and Holowaychuk, 1970a, b; 1971).

In New Zealand, podzolisation and the development of spodosols is often associated with *Podocarpus* spp. and the Kauri pine.

PARENT MATERIAL

It can be very difficult to assess the role of parent material in soil formation. Parent material often only provides the framework within which other factors and processes operate. As noted previously, Paton *et al.* (1995) prefer the term 'lithospheric material'. The term 'parent material' tends to imply a static situation, whereas 'lithospheric material' conveys the impression of a more dynamic situation with soil formation occurring on and being influenced by depositional material. However, as 'parent material' occurs as a term in all the major soil models it will be used here, but it should be understood that the wider nature of parent material is implied.

Parent material influences soil properties to varying degrees. Birkeland (1984) has stressed that the influence appears to be greatest in the early stages of soil formation and in drier regions. With time, other soil-forming factors, such as climate and biota, overwhelm the influence of parent material. The length of time that soil has been forming will also be significant. However, there are a number of what can be called 'extreme cases' where some rock types and materials can control the evolution of soils to a great extent. Podzolisation takes place best on acid parent materials poor in bases, such as sandy and/or crystalline materials. Rocks with a high carbonate content inhibit podzolisation because a high base content keeps pH high and prevents the translocation of Fe and Al. Thus soils formed on limestones would be another of these extreme examples, being commonly thin, fine-grained and with a sharp contact with bedrock. The soil tends to be the insoluble residue left following dissolution of the carbonate. However, there is a growing body of evidence to suggest that some of this material is of aeolian origin. Nevertheless, the general nature of such soils is the result of the specific characteristics of limestones and the way they weather.

Volcanic air fall deposits, such as ash and pumice, also tend to produce distinctive soils (Shoji *et al.*, 1993), so much so that they are given their own soil order, andosols (andisols). Common properties of such soils are low bulk density (<0.859 g cm^{-3}) and a high content of alteration materials of low crystallinity. Soils on such volcanic materials can form extremely quickly, the rate depending on the size of ejected material. Coarser particles develop into soils at a slower rate than finer particles.

Many volcanic rocks are rich in feldspar and ferromagnesian minerals, which weather to provide abundant oxides and hydroxides of iron and aluminium. Soils developed on lava also exhibit distinct characteristics. Gibbs (1980) has recognised three soil types, Kiripaka, Ruatangata and Okaihau, developed on massive basaltic lava in New Zealand. Kiripaka soils are the youngest and are moderately leached, with 30–50 per cent clay, mainly kaolinite and gibbsite, and small amounts of micas, vermiculite and amorphous oxides. Older soils, such as the Ruatangata soil, are moderately to strongly acid, well leached of cations and containing 50–80 per cent, kaolinite and gibbsite. The oldest soil, the Okaihau, is thoroughly leached, containing 30–50 per cent clay, mostly gibbsite, with some kaolinite and small nodules of iron, aluminium and manganese.

Parent material influences soil properties through the process of weathering and then through the influence of the weathered material on subsequent soil processes. Rock types influence the weathering process through their varying physical and chemical properties. Some rock types are more susceptible to weathering than others. Although physical weathering will be important in the early stage of soil formation, it is chemical weathering that has the greatest influence. The rate of mineral weathering depends on chemical composition, crystal shape and size, and degree of crystal perfection. Weathering rate is governed by available surface area, therefore large minerals are harder to weather than several small minerals occupying the same volume.

Knowledge of the weatherability or stability of various minerals is based largely on the work of Goldich (1938), who proposed a stability series that has now become well established (Table 4.4). For granite, the order becomes:

plagioclase > biotite > orthoclase > muscovite > quartz

However, the general validity of this order is now being challenged. Wilson *et al.* (1971), from their study in northeast Scotland, suggested an order of :

Plagioclase = orthoclase > muscovite > biotite > microcline > quartz

Relative stabilities cannot be considered invariant but must be assessed in terms of the prevailing weathering conditions.

The discovery of a relatively consistent stability series has prompted much speculation as to why

Table 4.4 Weathering sequence for common rock-forming minerals

INCREASING STABILITY		
olivine		Ca^{2+} plagioclase
augite		
	hornblende	
biotite		Na^{+} plagioclase
	K^{+} feldspar	
	muscovite	
	quartz	

Source: After Goldich (1938)

different minerals respond in the way they do. Ollier (1984) has suggested that the series may be related to basicity. The more cations that can be replaced by hydrogen, the more weatherable a mineral is. This relates to the fact that the stability series is generally the reverse of Bowen's reaction series, which lists minerals in their order of crystallisation. The minerals formed first in igneous rocks tend to use up most of the bases. Thus rocks that crystallise more slowly tend to be more acidic and more resistant to weathering. Other factors appear to be the bonding energies between cation–oxygen bonds, the molecular structure and the degree of packing of the molecules. The different weatherability of biotite mica and muscovite may be due to the difference in orientation of hydroxyl ions with respect to the plane of the mica. As was seen in Chapter 2, the alterability of silicates depends on the abundance of ions of large diameter, small charge and weak ionic potential in bulky polyhedra with relatively weak bonds. The cations in orthoclase feldspar are large whereas those in plagioclase feldspar are small, which may explain their differing susceptibilities to weathering.

Quartz is very resistant to chemical weathering but will dissolve under certain conditions. Feldspars are almost as resistant as quartz but possess a well-developed cleavage, which enables rapid alteration to take place. The pyroxene group of minerals, which includes augite, possess good cleavage, which aids rapid alteration. Amphiboles are more resistant to weathering, with hornblende being the most resistant. The mica group of minerals possess a sheet-structure crystal lattice, which enables the grains to break down in flakes. Muscovite is more stable than biotite. Zircon is one of the more stable minerals and has often been used as a standard against which weathering changes can be assessed (see Chapter 3 and Box 3.1).

The ease with which chemical weathering occurs will also depend on the relative mobilities of the constituent ions. Chemical elements such as Ca, Mg, Na and K are removed from rocks and weathered material more rapidly than others. This has been called chemical winnowing. Anderson and Hawkes

(1958) found that the order of mobility in granitic and schistose rocks from New Hampshire was:

$$Mg > Ca > Na > K > Si > Al = Fe$$

Polynov (1937) suggested that the order was:

$$Ca > Na > Mg > K > Si > Al = Fe$$

and Miller (1961) found that mobilities in different rock types in New Mexico were:

Granite $Ca > Mg > Na > Ba > K > Si > Fe = Mn$ $> Ti > Al$
Quartzite $Ca > Na > K > Mg > Fe > S > Al$
Sandstone $Ca > Na > K > Si = Al$

Summarising the available information, the following relative mobilities in order of decreasing mobility would be:

$$(Ca, Mg, Na), K, Fe, Si, Ti, Al$$

The main external factors that appear to influence this mobility are leaching, pH, E_h, fixation, and retardation and chelation.

A knowledge of rock mineralogy allows rocks to be placed in their order of susceptibility to chemical weathering. The rank for igneous rocks, in order of decreasing resistance, appears to be granite > syenite > diorite > gabbro > basalt (Gerrard, 1988). For temperate North America, Birkeland (1984) has proposed the following sequence: quartzite, chert > granite, basalt > sandstone, siltstone > dolomite, limestone; and for weathering by salt crystal growth, Brunsden (1979) has suggested the sequence diorite > dolerite > granite > gneiss > shale > sandstone > limestone > chalk.

If weathering has been intense and the landscape stable, considerable thicknesses of weathered material, known as the weathering profile, may develop with specific characteristics related to the nature of the rock being weathered. The weathering profile will vary considerably from place to place because of variations in rock type and structure, topography, rates of erosion, groundwater conditions and variations in climate. But, especially in the early stages, the nature of the weathering profile will be determined by the parent material and the way it weathers. The *in situ* weathered rock is usually called **saprolite** and is recognised by the presence of undisturbed joint planes, veins or similar rock structures. As was noted previously, the ability to recognise such indications of rock structure is one of the ways of differentiating C from B horizons.

Weathering profiles have been examined most thoroughly on igneous rocks and in tropical or subtropical regions, where weathering has been sufficiently intense to produce thick profiles. Therefore most tropical soils are developed on weathering profiles. Most descriptions of weathering profiles follow the scheme devised by Ruxton and Berry (1957) for granites in Hong Kong (Figure 4.9). A description of the sequence is shown in Box 4.2. These zones were characterised on the basis of the chemico-mineralogical changes that have taken place and the state of physical disintegration. Weathering profiles on sedimentary rocks can be extremely variable, but the general principles still

apply. Most weathering profiles exhibit trends of decreasing particle size and increased clay content towards the surface. Profiles on carbonate rocks are usually just the insoluble portion of the rock, such as quartz, chert, iron and manganese oxides, and some clay minerals. Deep solution cavities, filled with soft clays, are common. Weathered residues on chalk are usually thin because of the rock's purity, but zones can usually still be recognised (Box 4.3). Weathering profiles on clay shales (Box 4.4) can be quite different from those on other rocks.

Weathering zone

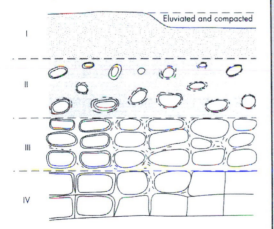

Figure 4.9 Weathering profile on granite rocks.
Source: After Ruxton and Berry (1957)

Box 4.2

GRANITE WEATHERING PROFILES

I   Residual debris composed of structureless sandy clay or clay sand, 1–25 m thick with up to 30 per cent clay, predominantly quartz and kaolinite, reddish brown when clayey and light brown or orange when less clayey.

IIa Residual debris with subordinate amounts of free, rounded corestones (less than 10 per cent); less than 5 per cent clay but plenty of clay-forming minerals; generally light in colour and with less than 10 per cent solid rock.

IIb As for IIa but with 10–50 per cent corestones and much of the original rock structure still preserved.

III Corestones are dominant, rectangular and locked, set in a matrix of residual debris; 50–90 per cent is solid rock.

IV   Partially weathered rock with minor amounts of residual debris along major structural planes, greater than 90 per cent solid rock although there may be significant iron staining and decomposition of biotite.

Source: After Ruxton and Berry (1957)

Box 4.3

WEATHERING PROFILE FOR CHALK

IV A structureless melange with unweathered and partially weathered angular chalk blocks set on a matix of deeply weathered remoulded chalk. Bedding and jointing are absent.

III Friable to rubbly chalk with joints closely spaced, ranging from 10 to 60 mm apart.

II Rubbly to blocky chalk with joints 60–200 mm apart. Joints are sometimes open, with secondary staining and fragmentary infilling.

I Unweathered medium to hard chalk, with widely spaced closed joints.

Source: After Ward *et al*. (1968)

Box 4.4

WEATHERING PROFILE ON SHALES

1 A surface zone of complete disintegration affected by freezing, temperature changes, wetting and drying, and chemical action.

2 A zone of advanced disintegration, subject to cyclic stresses due to groundwater fluctuations. This zone possesses numerous cracks and is softer and has a higher water content than the zone below.

3 A zone of medium disintegration subject to deep-seated strains, which may be due to release of strain energy.

4 Unweathered shale.

Source: After Bjerrum (1967)

It was stressed earlier that many soils develop in transported materials, and it will be the characteristics of such transported materials that will influence soil characteristics. River floodplains are classic situations where close relationships exist between the nature of deposited materials and the soils developed on them, but other good examples would include glacial materials and some coastal depositional areas (see Gerrard, 1992a, for a fuller analysis). Alluvial materials can be extremely variable, depending on river flow and flood frequency and the types of material being carried by the river. However, patterns of materials and soils can be identified, as was shown in the Indus valley by Holmes and Western (1969) (Box 4.5).

TOPOGRAPHY

A close examination of the various state factor equations, discussed earlier, will reveal that 'relief' and 'topography' are often used in a very general, perhaps even confusing, way. Huggett (1995) has pointed to this confusion. 'Topography' generally refers to the configuration of the land surface and is comparatively unambiguous. However, 'relief' is often used very loosely. It is sometimes used synonymously with topography and also to mean gradient angle. In its most restricted sense, 'relief' is the vertical difference between the highest and lowest elevations in a region. In this section, the term 'topography' will be used to avoid some of this confusion. Where necessary, more specific aspects of topography, such as slope steepness or gradient angle, will be used.

Topography influences the way in which soils develop by controlling the operation of near-surface processes. Thus it is one of the determinative factors as defined by Paton *et al*. (1995). Topography exerts its influence by a combination of gradient angle, slope form and position, i.e. whether soils develop on lower, mid-slope or upper slope positions. The influence of topography is implicit in the development of soil catenas (see Chapter 7), which are groupings of soils linked in their occurrence by

Box 4.5

RELATIONSHIP BETWEEN TEXTURE AND LANDFORMS IN THE INDUS VALLEY

Percentage sites per textual group

Landform association	AL	A	Bd	B	Bs	Xlm	X	Xhv	Cs	C	Cd	Cdv	D	Dv	Total sample
Low bar deposits	41	27.5	0.5	4	–	4	2	–	14.5	6.5	–	–	–	–	200
Channel scar deposits	26.7	26.7	–	6.7	–	6.7	33.3	–	–	–	–	–	–	–	15
Channel infill deposits	0.7	–	–	–	0.7	–	1.3	5.7	2.5	24.8	9.6	–	35.0	19.7	157
Levee deposits	10.2	21.3	5.5	33.1	7.1	12.7	21.6	–	0.8	–	–	–	–	0.8	127
Shallow cover floodplain	0.4	1.8	–	0.4	–	0.4	10.2	11.1	5.8	53.3	8.9	–	6.7	0.9	225
Deep cover floodplain	–	–	0.5	1.0	1.0	–	–	0.5	–	0.5	8.3	–	80.7	7.3	192
Backswamp deposits	–	–	–	0.6	–	–	0.3	1.5	–	2.4	0.3	5.4	6.3	83.3	336

Key to textual groups

Nomenclature	*Characteristic texture*
AL	predominantly coarse sand and loamy sand
A	mixture of sands and loams
Bd	sands and loams overlie silts and clays at depth >100 cm
B	sands and loams overlie silts and clays at depths between 50 and 100 cm
Bs	sands and loams overlie silts and clays at depths <50 cm
Xlm	complex but basically sands and loams
X	complex sequences of fine and coarse textures
Xhv	complex but basically silts and clays
Cs	silts and clays overlie sands and loams at depths <50 cm
C	silts and clays overlie sands and loams at depths between 50 and 100 cm
Cd	silts and clays overlie sands and loams at depths >100 cm
Cdv	upper horizons dominated by clay
D	almost entirely silts and clays
Dv	clays probably dominant

Source: From Holmes and Western (1969)

conditions of topography and are repeated in the same relationships to each other wherever the same conditions are met.

Emphasis is usually placed on the difference between freely drained upper parts of slopes and poorly drained lower portions. Slope steepness is an important factor, as steeper angles reduce the amount of water infiltrating and percolating through the soil and increase the removal, sometimes through accelerated erosion, of the upper portions of the soil profile. Soil and water can and do move downslope. Water moves as overland flow (surface wash) and throughflow, and material can move with this water or by creep and more rapid forms of mass movement. Although these processes act universally, they can be influenced by climate.

The major pathways by which water moves across and through soil on a slope are shown in Figure 4.10. The essence of Hortonian overland flow is that, if rainfall intensity is greater than the infiltration capacity of the soil, overland flow will occur. Hortonian overland flow will develop from the upper slopes to the lower slopes, and at a critical distance downslope the depth of flow will become sufficient to start eroding and transporting soil material.

The way in which slope form influences the operation of these processes has been shown by Hack

and Goodlett (1960). They classified slopes within a drainage basin and noted the different relationships with runoff. Runoff on convex spurs or nose slopes was proportional to a function of the radius of curvature of the contours. On relatively linear side slopes, runoff was proportional to a linear function of slope length and in the concave hollows was proportional to a power function of slope length.

It is now generally recognised that, in most environments with a well-developed soil and vegetation cover, the movement of water downslope within the soil is more important than overland flow. This means that the properties of the soil assume a greater significance and may be more important than basin morphometry in controlling runoff rates. Water moves through the soil in a downslope direction, implying that if saturated soil conditions are to occur, they will occur at the slope base. During a rainstorm, the saturated area may intersect the ground surface over the lower part of the slope and overland flow could occur. As the storm continues, the saturated wedge increases upslope and the amount of surface flow also increases. Kirkby and Chorley (1967) suggest that maximum water flow would occur at the base of the slope, in hollows, in slope profile concavities and in areas of thin and permeable soils. Dunne (1978) has shown that the recurrence interval of runoff-producing events is of the order of 10^2 to 10^3 years on straight, well-drained hillsides, whereas it is 10^{-1} years in the lower parts of shallow, moist swales and 10^1 to 10^2 years on the lower concave portions of well-drained hillsides. One of the most important factors is the thickness of soils on different parts of a slope. In some cases, simple relationships exist between soil properties, slope position and factors such as infiltration capacity and moisture storage (Table 4.5).

Water moving across slopes sometimes carries material with it. Thus most models of surface water erosion, such as the universal soil loss equation (see Chapter 8), include a topographic effect. Three main components of topography affect erosion by surface wash: slope steepness, slope length and slope shape. Erosion potential is greater on long slopes because

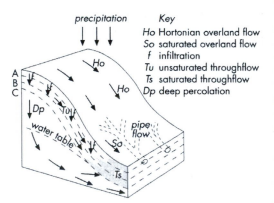

Figure 4.10 Various ways in which water can move downslope.

Table 4.5 Relations between landscape components and some soil moisture characteristics

	Watershed area (%)	Average slope (%)	Storage potential (cm)	Infiltration rates (cm h⁻¹)	
				initial	final
Upland soils	44.4	35.6	8.6	14.66	0.58
Hillside soils	46.7	12.7	3.8	5.03	0.25
Bottomland soils	8.9	61.0	14.2	28.88	0.58

Source: After England and Holtan (1969)

of a downslope increase in surface flow depth. Several empirical relationships have been established relating soil transport by surface wash to slope length and slope gradient, some of which are shown in Box 4.6. Other measurements have shown that soil loss and slope angle are related by power functions ranging from 0.7 to 2, with most of the values falling between 1.35 and 1.5.

Box 4.6

SOME EMPIRICAL RELATIONSHIPS BETWEEN SOIL TRANSPORT AND SLOPE LENGTH AND SLOPE ANGLE

Zingg (1940): $S \propto x^{1.6} \tan^{1.4} \beta$

Musgrave (1947): $S \propto x^{1.35} \tan^{1.35} \beta$

Kirkby (1969): $S \propto x^{1.73} \tan^{1.35} \beta$

where S = soil transport (in $cm^3 cm^{-1} yr^{-1}$)
        x = slope length (in metres)
        β = slope gradient

Marked differences in soil erosion rates occur on slopes with different surface form. This was shown in an analysis of soil gains and losses on different landform elements in Saskatoon, Canada, by Pennock and de Jong (1980) (Figure 4.11). Erosion on convex slopes increases rapidly as the slope steepness and slope length increase downslope. Concave slopes tend to counteract the increased downslope potential of greater runoff by gentler angles. Slope gradient and runoff potential are in opposition. In a study by Meyer and Kramer (1969), the maximum eroded depth was least on concave slopes, followed by uniform slopes, complex slopes and convex slopes. However, Young and Mutchler (1969) found that convex slopes had less soil eroded than expected because convexities were concentrated on slope crests with less available water.

In an interesting study in Poland, Jahn (1963) found that on the steep upper parts of convex–concave slopes the soil was so severely eroded that unweathered loess was exposed on the surface. The lower slopes were covered in slope wash material. This study also emphasised the distinction between concentrated and unconcentrated wash. Line degradation (rill erosion) was greatest on concave slope sections, but so too was accumulation by undifferentiated slope wash. The threshold between the two sets of processes is clearly of great importance.

Many slopes and their soils are affected by mass movements in one form or another. As mass movements in general are discrete events, they tend not to be integrated with the entire slope. However, it is possible to establish general relationships between type of mass movement and factors of topography such as slope angle and slope height (Figure 4.12).

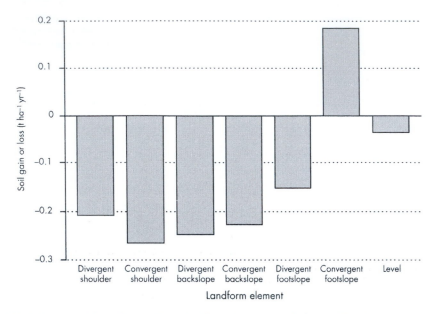

Figure 4.11 Soil gains and losses from different landform elements near Saskatoon, Canada.
Source: Based on Pennock and de Jong (1987)

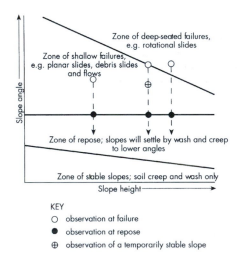

KEY

○ observation at failure
● observation at repose
⊕ observation of a temporarily stable slope

Figure 4.12 Differentiation of mass movements on the basis of height and angle of failure.
Source: After Skempton (1953)

As mentioned in Chapter 1, slopes and soils tend to act as integrated systems, therefore it is not surprising that good relationships can be established between topographic factors and soil properties. One of the earliest studies to establish such relationships was that by Norton and Smith (1930) on loess soils in Illinois. They found correlations between slope and soil structure, texture and consistency and an inverse relationship between slope angle and depth of the textural B horizon.

A number of studies (e.g. Furley, 1968; 1971; Whitfield and Furley, 1971) have demonstrated that there may be fundamental differences in relationships on convex and straight slope sections and on concavities. High correlations were found between soils and slope angle on convex elements and maximum segments, but relationships on concave elements were poorer. Generalised relationships for the convex and maximum slope portions are shown in Figure 4.13. Acidic soils showed a decrease in pH with increasing slope gradient, whereas the

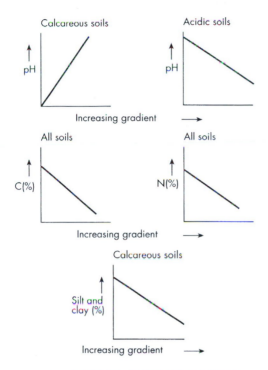

Figure 4.13 Generalised relationships between soil properties and slope angle.
Source: From Furley (1968)

Table 4.6 Dominant factor explaining the variation in soil properties on chalk slopes

Dominant factor	Total slope	Upper slope	Lower slope
Slope angle	7	21	10
Slope position	34	21	20
Angle and position equal	7	6	17

Source: From Furley (1971)

relationship was reversed for calcareous soils. All soils, acidic and calcareous, exhibited inverse relationships between gradient and carbon and nitrogen content and, for calcareous soils, percentage silt and clay declined with increasing slope angle.

Many of the relationships just discussed are likely to have been influenced by the position of the soil on the slope. On simple convex–concave slopes, slope gradient is highly correlated with position. Of the forty-eight slopes analysed by Furley (1971), on only seven was slope angle the dominant factor in explaining variations in soil properties over the entire slope. Slope position was the dominant factor on thirty-four of the slopes (Table 4.6). When the slopes were subdivided into upper and lower portions, interesting changes in the soil relationships occurred. For upper slope portions, slope position

and slope angle were equally dominant, but for lower slopes, slope position was the major factor. Relationships were also stronger between slope position and soil properties on the lower slopes. Using these results, Furley (*ibid.*) has produced a simple model to describe and explain the distribution of soil properties on slopes (Figure 4.14). The basis of the model is that soluble minerals and exchangeable ions are leached from upper slopes, move downslope and are deposited in lower slope positions. The greatest concentrations of soil properties, such as organic content, would be expected on the gentle gradients at the tops and bottoms of slopes. The different relationships on upper and lower slopes appear to reflect different processes: upper slopes are influenced by processes of erosion and transport, whereas lower slopes are ones of deposition and transport.

TIME

Most soil properties exhibit quite a rapid change in values at the commencement of soil formation, but the rate of change soon declines and gradually slows down until an approximation of a steady state is reached. Steady state appears to be reached more quickly by properties of the A horizon than by properties of the B horizon, and steady state is reached probably more rapidly with organic matter properties than with any other soil property. Even so, time to reach steady state may range from 200

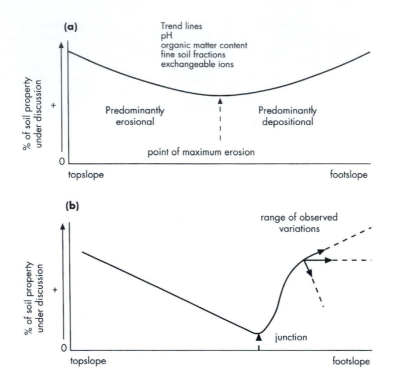

Figure 4.14 (a) Theoretical distribution of soil properties with slope form. (b) The distribution of certain soil properties on chalk slopes.
Source: From Furley (1971)

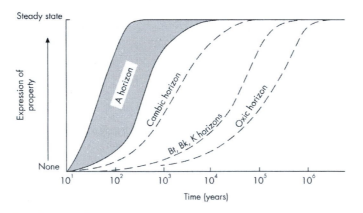

Figure 4.15 Rate of change of certain soil properties with time.
Source: From Birkeland (1984)

to 10,000 years (Figure 4.15). Data also suggest that, if conditions change, the A horizon can reach a new steady state quite quickly.

Studies of soils on recent glacial moraines have established that, in general, the pH of the top soil layers is reduced, calcium carbonate is leached out of the system, nitrogen content decreases and organic matter increases with age. Other soil properties (such as base saturation) will change as a result of these trends. The results also show that vegetation type influences the speed of change. In the glacial moraine studies, the greatest change per unit time occurred under alder.

Other properties of the A horizon develop almost as rapidly as organic matter constituents, probably in response to changing organic matter. Soil pH declines quickly and probably reaches a steady state as quickly as organic matter. Nitrogenous matter also changes rapidly in the first 1,000 years. Carbonates can be leached rapidly, but this will depend on local climates. If it can be shown that calcium carbonate was distributed uniformly in soil parent material, it is possible to calculate how long it might take to redistribute the calcium carbonate by solution, translocation and re-precipitation. A model to achieve this has been provided by Arkley (1963), based on the volume of water passing various levels in the soil as indicated by climatic data and information on calcium carbonate solubility. One calculation produced a figure of 9,800 years to remove carbonate from the uppermost 58 cm and then precipitate it in the 58 to 165 cm depth interval, based on 3.3 litres of water moving through every square centimetre of soil per 1,000 years. In a chernozem in former Yugoslavia, 35 per cent of the carbonate has been lost from the uppermost soil during the last 10,000 years, giving a rate of removal of approximately $0.035 \, \text{mg} \, \text{g}^{-1} \, \text{yr}^{-1}$ (FitzPatrick, 1983).

The properties of the B horizon appear to change and develop more slowly. The development of an argillic B horizon (Bt) has attracted much attention (see Chapter 5 for horizon development). Increased clay content in the B horizon can be achieved in a number of ways, such as weathering, aeolian influx,

clay formation and translocation. These are all inherently slow processes. It is difficult to obtain accurate information, because the clay content of the B horizon has to be compared with the rest of the soil profile to establish that clay enrichment has occurred. Therefore a clay enrichment index, which is the clay content in the Bt horizon minus that in the C horizon multiplied by the thickness of the Bt horizon, is sometimes used.

Clay content with depth relationships are often compared for soils on landforms of known age to try to assess clay enrichment with age, but there may be considerable variations depending on climate and parent material and especially aeolian influx. It has been noted that Bt horizons formed in about 350 years near Lubbock, Texas (Birkeland, 1984), but this is probably an exceptional rate owing much to a significant mud rain input.

In the Nevada desert, a Bt horizon has formed in a soil that might be less than 2,000 years old (Peterson, 1980), but this is attributed to both a dust input and the dispersive effects of Na^+. In contrast, in the cold desert of Antarctica, with less than 1 cm of annual water equivalent, Bt horizons are not present in soils several million years old. The maximum time for the development of Bt horizons has been estimated as 40,000 years in the San Joaquin valley, California, and 140,000 years for the Sierra Nevada and Rocky Mountains. The greater length of time for the last two locations seems to reflect low moisture status and low dust influx.

Many soils appear to become redder with age, and much attention has been paid to rubefication in the B horizon. Generally, red soils form fairly rapidly in areas with high temperatures, therefore redness in soils is probably a result of a combination of age and climate. The rate of increase in redness in cold, dry arctic conditions is very slow. The redness, in some cases, may reflect palaeoclimates and may not reflect soil age. Harden (1982) has devised a dry rubefication index to assess the development of redness. Changes in three colour indices with time are shown in Figure 4.16.

Amounts of free iron and aluminium and phosphorus fractions in soils are very clearly time-

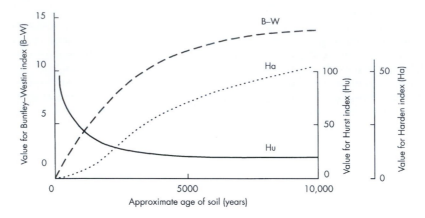

Figure 4.16 Development of the weighted colour index values with time for a sequence of Holocene soils in the Wind River Mountains, Wyoming.

Source: From Miller and Birkeland (1974); Birkeland (1984)

dependent. There is a general trend of accumulation of iron and aluminium and depletion of phosphorus. Greatest accumulations of iron and aluminium and greatest depletion of phosphorus occur in relatively wet and warm environments, whereas soils in cold, dry arctic environments commonly show no trends that can be referred to soil formation. In New Zealand, trends in phosphorus fraction depletion are most noticeable in areas with the highest precipitation (Walker and Syers, 1976).

A great amount of information on the way soils develop and the way in which specific soil properties vary with time has been obtained from an analysis of soil chronosequences. A chronosequence can be defined as a generally related suite of soils in which vegetation, topography and climate are similar (Harden, 1982). This is the basic definition for the individual state factor equation examined earlier. However, the use of soil chronosequences is not without its problems. When a soil property is plotted against soil age, changes with age may or may not be the result of a single pedogenic process acting on soils through time. Processes or changes in conditions in the history of a soil are not always recorded or preserved in morphology. However,

with care, much useful information can be obtained from a study of chronosequences.

Chronosequences are usually obtained by examining soils on landforms of different ages in a region. Glacial moraines and river terraces, which can often be placed in age sequences and which can sometimes be dated quite accurately, are especially useful in this respect. Mellor (1985) has examined chronosequences on neoglacial moraine ridges in Jostedalsbreen and Jotunheim, southern Norway. Some of these results are shown in Figure 4.17. Organic-rich surface horizons increase significantly with age, although extrapolation of the data suggests a decline in the rate of increase after about 230 years. There is a significant increase in the ratio of dithionite-extractable iron (Fe_d) in the B horizon compared with in the C horizon at Austerdalsbreen. The Fe_d data show a tendency towards a steady state within 230 years. Significant increases in CEC and exchangeable potassium and magnesium with age were noted in the bleached horizon. There was no evidence of clay increase with depth, and time-related trends in particle size were not observed, but there was an age-related increase in the thickness of the E horizon. This reflects the operation of

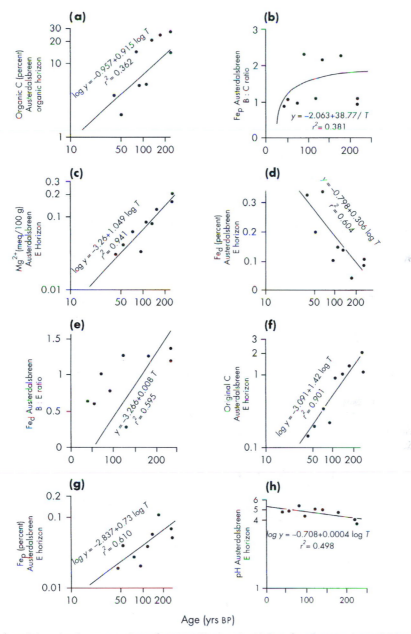

Figure 4.17 Selected soil property chronofunctions illustrated with best-fit equations, Austerdalsbreen, Norway.
Source: From Mellor (1985)

translocatory processes with organic carbon and pyrophosphate-extractable iron and aluminium moving in, but Fe_d was significantly depleted from this horizon but enhanced in the B horizon.

A useful summary of thirty-two published chronosequences has been provided by Bockheim (1980). He correlated each soil property with time using a variety of regression models. These chronosequences varied in parent material and occurred in a variety of climatic environments, enabling the influence of these factors to be assessed. His basic conclusions were as follows (see also Birkeland, 1984: p. 223):

1   rates of decrease in pH and in base saturation are similar and do not appear to be related to parent material and climate;
2   rates of increase in B horizon clay content and in soil thickness are particularly related to the clay content of the parent material;
3   rates of increase in soil thickness, depth of oxidation and B horizon clay content are positively correlated with mean annual temperature; and
4   rate of change in C:N ratio does not appear to be correlated with either climate or parent material.

So far, temporal changes have been examined with respect to single soil properties, but clearly there are general progressions with time for the entire soil profile. Trends in profile development indices with time, which have provided useful information, have also been established (Birkeland, 1990). Profiles tend to become more differentiated with time. This has been shown very clearly in soils on a sequence of river terraces in the Southern Alps, New Zealand (Tonkin and Basher, 1990). At one locality, Slovens Creek, soils on the youngest terrace (<1,000 years) possess simple A/2BC profiles with a sandy loam A horizon formed in alluvium. Soils on older terraces (10,000–20,000 years) possess A/AB/Bw/BC/2C profiles. Morphological development involves increasing B horizon rubefication, accompanied by clay content increases from 28 to 36 per cent. Younger soils have a higher pH, a lower percentage

of carbon and higher total amounts of exchangeable bases. Increasing complexity of soil profiles is well seen in desert soils in New Mexico (Box 4.7).

There is also a good correlation between soil orders and age (Figure 4.18). Oxisols seem to require a long time to develop, whereas spodosols may develop in a few thousand years. Soils with argillic horizons, such as mollisols and aridisols, would require longer periods of time (about 10^4 years) to form. Little-developed soils, such as entisols, would be classed as such almost as soon as pedogenesis commenced. Histosols and vertisols could also develop in a comparatively short period of time.

The question of a steady state has been raised a number of times in this section, and it has been assumed that many soil properties reach a steady-state condition. There is no doubt that the rate of change of soil properties declines quite rapidly with time. But do they eventually reach a steady state? Soil profiles can only be thought of as being in a steady state when most of their diagnostic properties are in a steady state. Also, the steady-state condition will be reached for different properties at different times as a result of the operation of different processes. Some will involve additions, others translocations or depletions. Recent data have shown that some soil curves have not flattened even after pedogenesis lasting 10^6 years. Plots of the Harden profile development index for central California have not flattened at the age of the oldest soil (3×10^6 years) (Harden, 1982). It is also very doubtful if soil-forming processes are constant for such lengthy periods. Steady state in soils may be one of those concepts that is rarely attained.

SUMMARY

Traditionally, the pathways of soil development have been related to a number of key controls, or state factors as they have been called. Models of soil development have been produced using these factors. Although there is little doubt concerning the influence of these factors, the nature of their

Box 4.7

SOIL AGE AND CHARACTERISTICS IN NEW MEXICO

Soil age in years	Features of soil development
100 years	Thin grey A horizon, vesicular in places, slight organic matter accumulation, original sedimentary bedding still present.
100–1,000 years	Very slight carbonate accumulation. Most of the original bedding has been destroyed, and only thicker beds of fine earth remain.
1,100–2,100 years	No distinct signs of carbonate accumulation in soil with little gravel, but in gravelly soils a weak but a distinct horizon of carbonate accumulation, chiefly in the form of pebble coatings.
2,200–4,600 years	In non-gravel soils, a weak carbonate horizon has developed in the form of coatings on structure units, and a distinct structure has developed.
Late Pleistocene	Oriented clay coatings on grains in a B horizon with clay enrichment, and a K horizon. A Km horizon has developed with a single laminar horizon on its upper edge.
Mid-Pleistocene	The same, with a distinct Km horizon with two or more laminar horizons (K21m) on top.

Source: From Gile and Hawley (1968)

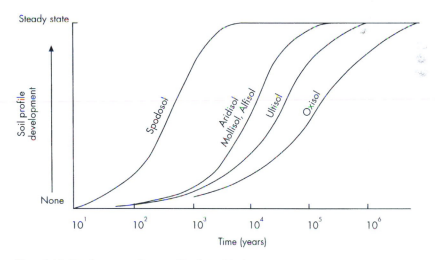

Figure 4.18 Development of some soil orders with time.
Source: From Birkeland (1984)

specific influence is increasingly being questioned, and it is rarely easy to isolate one factor among the many as being the most important in any specific situation.

ESSAY QUESTIONS

1 **Examine the rationale behind state factor equations.**

2 **Why is it so difficult to assess the relationships between parent material and soil type?**

3 **Examine the problems in establishing EITHER soil climofunctions OR soil chronosequences.**

FURTHER READING

Birkeland, P.W. (1984) *Soils and Geomorphology*, New York: Oxford University Press.

FitzPatrick, E.A. (1986) *An Introduction to Soil Science*, Harlow: Longman.

Gerrard, A.J. (1992) *Soil Geomorphology*, London: Chapman & Hall.

Jenny, H. (1941) *Factors of Soil Formation. A System of Quantitative Pedology*, New York: McGraw-Hill.

Paton, T.R. (1978) *The Formation of Soil Material*, London: George Allen & Unwin.

Paton, T.R., Humphreys, G.S. and Mitchell, P.B. (1995) *Soils: A New Global View*, London: UCL Press.

Tedrow, J.C.F. (1977) *Soils of the Polar Landscapes*, New Brunswick, NJ: Rutgers University Press.

5

HORIZONS AND SOIL CLASSIFICATIONS

SOIL HORIZONS

Soil is not a random assemblage of organic and inorganic particles. The soil processes outlined in the previous chapter may impart a definite vertical organisation to soil. However, this organisation is not necessarily present in all soils. This vertical organisation usually manifests itself in the form of distinct horizons that differ in their physical, chemical and biological attributes. Horizons reflect the operation of soil processes and, to some extent, the controls on those processes examined in the previous chapter. Thus it is not surprising that many soil classification schemes have relied on the recognition of the type, number and sequence of horizons. Horizon recognition and the nomenclature used to describe those horizons has become a kind of 'shorthand' that enables rapid communication about the morphology, possible genesis and sometimes classification of a soil. However, many difficulties are involved in horizon recognition, prompting FitzPatrick (1983) to state that horizon recognition has been more of an art based on experience rather than a science based on any set of defined principles. In some soils, horizons are clear and relatively unambiguous, but in many soils, laboratory work is needed to define horizons in respect of the

characteristics needed to classify that soil. This point is taken up in the section on soil classification. It has been stressed many times that soils function as systems and therefore horizons should not be examined in isolation, and it will be necessary to compare properties of one horizon with those above and below it.

The change from one horizon to another varies in degree of sharpness and in outline. Colour is usually the most obvious change from one horizon to another, but structural and textural changes may also be clear. Classes of distinctiveness and outline have been produced by FitzPatrick (*ibid.*; see Box 5.1). Thin horizons generally possess sharp or very sharp boundaries, and boundaries tend to become less distinct as horizon thickness increases. Boundaries reflect the transfer and transformation processes operating in the soil.

When discussing horizons, it must be remembered that horizons will change laterally as well as vertically (Figure 5.1). A single horizon may remain unchanged, or it may thicken or thin laterally. It may also become deeper or shallower. It sometimes ends completely and a new horizon commences, or it may be replaced by another horizon starting from its base or top. The last possibility is that a horizon may undergo a gradual change in properties while

Box 5.1

HORIZON BOUNDARIES

Distinctness classes		**Outline classes**
Abrupt	– change takes place within 2 cm	Smooth – almost straight
Sharp	– change takes place within 2–5 cm	Wavy – gently undulating
Clear	– change takes place within 5–10 cm	Lobate – with regular lobes
Gradual	– change takes place within 10–20 cm	Irregular – strongly undulating and mamillated
Diffuse	– change takes place within >20 cm	Tongued – forming tongues into the underlying horizon, shallow
Source: From FitzPatrick (1983)		tongues and deep tongues.

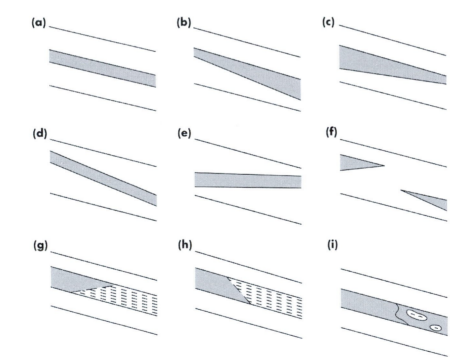

Figure 5.1 Possible horizon changes in a downslope direction.

retaining its identity and continuity. Changes will occur laterally on horizontal surfaces as a result of changes in some or all the factors affecting soil development, such as climate, biota or parent material, but changes are usually most noticeable as a result of topographic changes. Such changes have been embodied in the concept of the catena which is discussed in Chapter 7. On a slope, there may be zones where one or more horizons undergo gradual change and zones where rapid changes take place, leading to substantially modified horizons over short distances.

Horizon designation

Horizons may be described simply as upper, middle or lower, depending on their position in the soil profile. However, more specific horizon designations having genetic connotations are usually used. The Russian soil scientist V.V. Dokuchaev is credited with introducing the ABC nomenclature system. The A horizon was the first horizon from the surface coloured by humus. It was regarded as possessing a uniform character with regard to humus content, colour, structure and texture. The C horizon was parent material unmodified by soil-forming processes. The B horizon was the second from the surface and transitional from the bottom of the A to the top of the C. It was the zone through which the content of humus gradually decreased from its value in the A horizon to zero in the C horizon. To a large extent, as will be seen, this system is still used.

Horizons that are commonly recognised in a variety of soils are called 'master' horizons and are designated by capital letters. Horizons H, O, E and R have been added to the ABC system of Dokuchaev (Figure 5.2). These horizons are produced by the processes of additions, transfers, transformations and losses described in Chapter 3. In situations where the properties of two master horizons merge, this is indicated by the combination of two capital letters, such as AB, AE, BC, etc. Horizons that consist of intermingled parts characteristic of different master horizons are designated by two capital letters separated by a diagonal stroke, e.g. E/B, B/C.

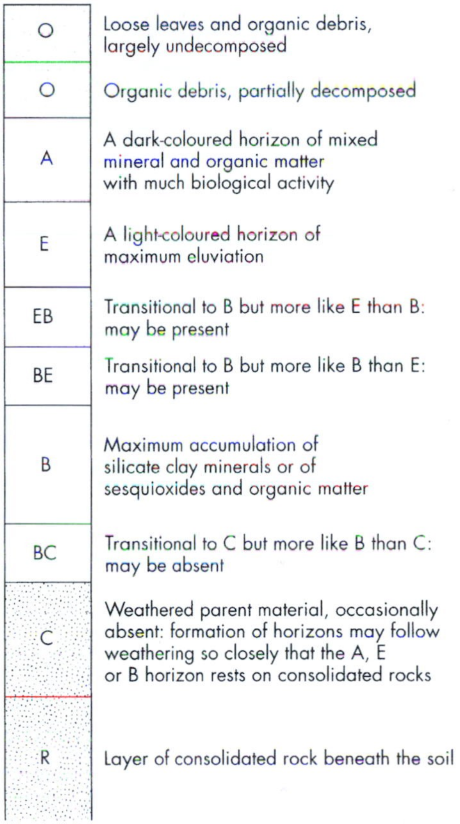

O	Loose leaves and organic debris, largely undecomposed
O	Organic debris, partially decomposed
A	A dark-coloured horizon of mixed mineral and organic matter with much biological activity
E	A light-coloured horizon of maximum eluviation
EB	Transitional to B but more like E than B: may be present
BE	Transitional to B but more like B than E: may be present
B	Maximum accumulation of silicate clay minerals or of sesquioxides and organic matter
BC	Transitional to C but more like B than C: may be absent
C	Weathered parent material, occasionally absent: formation of horizons may follow weathering so closely that the A, E or B horizon rests on consolidated rocks
R	Layer of consolidated rock beneath the soil

Figure 5.2 Master horizons.

Subdivisions of master or even transitional horizons are indicated by Arabic numbers following the horizon symbol, such as C1, C2, A1, A2, etc.

Horizon characteristics

Unaltered bedrock is termed the R layer. It is clearly not a soil horizon and in weathering terminology the junction between the R layer and the C horizon is referred to as the weathering front. As weathering progresses, the R layer is transformed into the mineral horizon of unconsolidated material, known as the C horizon, from which the soil generally develops. As the C horizon is little affected by biological activity, technically it is not part of the soil solum. It is affected by physical and chemical processes and may show evidence of calcium carbonate cementation and accumulations of silica and soluble salts. The C horizon may also be a superficial deposit overlying bedrock before weathering has commenced. If this material is thin, weathering may penetrate into the underlying bedrock and may result in a composite C horizon with unusual properties. The specific properties will depend on how different the overlying material is from bedrock.

At the surface, additions of organic material will produce a soil layer that will differ from the rest of the soil on account of its higher content of organic matter. Ellis and Mellor (1995) have provided a synthesis as to how the nature of the surface horizon depends on the balance between the processes of organic matter addition and its subsequent transformation, transfer and loss. If organic input greatly exceeds the rate at which it is transformed, transferred and mixed with the rest of the soil, a large organic accumulation will occur on the surface. This is generally referred to as an H horizon (Figure 5.3a), in which the original forms of the vegetable matter are usually recognisable with the unaided eye. Where the rate of organic matter transformation is slightly greater or where the rate of addition is slightly less, an organic-rich surface horizon still results, but different stages of organic matter de-

composition can be recognised. This is usually called an O horizon (Figure 5.3b), and the original forms of the vegetable matter are not usually visible with the unaided eye. Where organic addition exceeds transformation but where transfer and mixing occur with little loss, a mixed organic and mineral horizon develops, known as the A horizon (Figure 5.3c). If organic matter is transformed, transferred and lost quickly, there will be little or no surface organic horizon (Figure 5.4d).

The horizon that forms between the A and the C is the B horizon, and it is easy to see why Dokuchaev called it the transitional horizon. It is affected by transformations within the C horizon and especially by transfers from the A horizon. Transfers and transformations within the B horizon complete the wide range of processes that affect its characteristics. The B horizon can be defined as a horizon in which rock structure has been obliterated or is only faintly evident. The B and C horizons can often be differentiated on the basis of intensity of chemical weathering, as defined by the weathering indices noted in Chapter 3. These indices can also be used to distinguish soils of different ages. Ruhe (1956) employed both chemical and mineral weathering indices in a comparison of soils on stepped erosion surfaces in Iowa (Table 5.1). This not only shows the way in which indices vary with soil horizon but also demonstrates the difference in age of the soils. However, great care must be taken when using these indices. More than one particle size should be

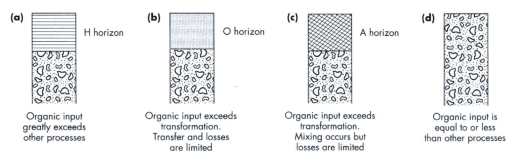

Figure 5.3 Development of surface horizons.
Source: After Ellis and Mellor (1995)

Table 5.1 Soil properties on stepped erosion surfaces in Iowa

Surface	Soil	Thickness of soil (in)	Thickness of B horizon (in)	Clay in B horizon (%)	Soil horizon	Wrh index	Wrl index
Recent	A	15	11	31.2	A	0.79	2.09
	B	32	23	32.2	B	0.92	2.13
	C	29	22	34.6	C	0.68	2.21
	average	25	19	32.3			
Late Sangamon	D	46	32	50.7	A	1.27	3.06
	E	70	56	49.1	B	1.12	2.49
	F	39	29	49.5	C	0.77	2.04
	average	52	39	49.7			
Yarmouth	G	87	70	51.4	A	2.11	4.85
	H	68	44	57.7	B	1.62	3.00
	I	85	62	50.7	C	1.28	2.57
	average	80	59	53.2			

Source: From Ruhe (1956)

examined, because mineral species content varies with particle size.

In general, B horizons are characterised by illuviation. Horizons created by eluvial processes are known as E horizons, defined on the basis of the removal of silicate clay, iron and aluminium, individually or in combination. Eluviation will usually create a change of colour if iron is removed, or a change of texture if clay is removed. Chemical characteristics and structure will also change. Eluviation and illuviation were examined in greater detail in Chapter 3.

Subsidiary characteristics of master horizons are usually designated by a suffix (Box 5.2). Thus a B horizon with a marked accumulation of illuvial clay will be a Bt horizon, and a B horizon with accumulations of oxides and hydroxides of iron and aluminium would be designated Bs. An A horizon disturbed by ploughing or other tillage activities will be Ap. Sometimes combinations of suffixes are used. B horizons with concretionary accumulations of calcium carbonate would be Bck, and if the accumulations were consolidated or indurated the designation would be Bmk.

Horizons form the basis of most soil classification schemes. The fact that they may be difficult to identify and to designate means that some classifications are not easy to operate. These problems are now explored by examining some of the more important soil classification schemes.

SOIL CLASSIFICATION

Why do we need to classify soils? One of the aims of any classification system is to organise knowledge. Classification also provides a framework for the storage and retrieval of information and enables communication. Thus Cline (1949) states that the purpose of any classification is to organise our knowledge so that the properties of objects may be remembered and their relationships understood most easily for a specific objective. McRae and Burnham (1976) have stated that, in general, there are five reasons for making a soil classification. The first reason is to create an impression of the nature of the soil profile or of the soil in an area in relation to others. Second, classifications simplify the

Box 5.2

MASTER HORIZON QUALIFICATIONS

b   Buried or bisequal horizon.
c   Concretionary accumulations; the nature of the concretionary material can be shown by an additional suffix, e.g. Bck (accumulation of calcium carbonate).
g   Mottling.
h   Accumulation of organic matter in mineral horizons.
k   Accumulation of calcium carbonate.
m   Strongly cemented, consolidated or indurated.
n   Accumulation of sodium.
p   Disturbed by ploughing or other tillage practices.
q   Accumulation of silica.
r   Strong reduction as a result of groundwater influence.
s   Accumulation of sesquioxides.
t   Illuvial accumulation of clay.
w   Alteration *in situ* as shown by clay content, structure or colour.
x   Occurrence of a fragipan.
y   Accumulation of gypsum.
z   Accumulation of salts more soluble than gypsum.

processing of soil data, and third, they can be used to reveal or study genetic relationships. Fourth, a classification may be devised as a mapping unit, and finally, a classification may be devised for a specific purpose such as for soil or land evaluation. Some of these aims are mutually irreconcilable, leading to a large number and variety of classifications. The basis of most soil classification schemes is taxonomic to try to understand soils and why they behave as they do.

Most taxonomic classifications rely on the identification of specific horizons that are regarded as being more important than others. These are known as diagnostic horizons and, as will be seen, lengthy descriptions have been provided in order to identify them. The complexity of these definitions often makes their use problematical: sometimes several diagnostic horizons are present in the same soil, which may make classification difficult. It also makes comparison with soils that contain only one of these horizons problematical. Some soils are classified on the lack of a specific diagnostic horizon. The fundamental point of soil classification is to produce groups in which individuals in any one group are generally similar to one another and different from those in other groups. A distinction can be made between monothetic and polythetic groups. In polythetic groups, individuals share many attributes, but no single attribute is either sufficient or necessary to confer class membership. In monothetic groups, the possession of one or more attributes is sufficient and necessary for class membership. It will be seen that most soil classifications involve both monothetic and polythetic groups.

Many classifications are hierarchical, with branching networks of soil types grouped into orders, classes and so on. The use of a hierarchical structure has been much criticised. Thus FitzPatrick (1983) has stressed that 'adopting this well trodden and fruitless path comes as a great surprise since it should be obvious that hierarchical systems have repeatedly failed to satisfy the demands of the soil continuum' (pp. 133–4), and 'the whole is a perfect example of creating a rigid structure without principles and their forcing the soil continuum into the various categories and at the same time having to produce the most complicated and peculiar definitions to accommodate the variability' (p. 134). This can result in some soil groupings containing a number of very divergent soils. Some of these problems will be examined with respect to specific soil classifications, which also tend to be national schemes.

Russian classifications

The early work of Russian soil scientists, especially Dokuchaev, his students and co-workers, has had a major influence on how soils are described and classified. Dokuchaev's classification used names for soil types that are still used today, although his classification scheme has been superseded. The basic elements of this scheme are shown in Table 5.2. A little later, the terms 'intrazonal' and 'azonal' were introduced to replace transitional and abnormal soils. Zonal soils possess well-developed pedo units and are formed under the influence of major vegetation and climatic types. Intrazonal soils possess well-developed pedo units but are formed under the influence of a specific local factor such as parent material or topography. Azonal soils are poorly developed because of a number of inhibiting factors such as slope steepness, erosion or deposition. Such a scheme appeared to work well in the great sweeps of gently undulating continental terrain characteristic of much of Russia and the former Soviet Union but has been found wanting in areas with greater variability of climate and topography and especially in areas that have experienced significant climatic changes in the Tertiary and Quaternary periods. Such a classification is also not best suited to old land surfaces such as are found extensively in Africa and Australia. Many of the terms and concepts remain in the system currently generally used in Russia (Rozov and Ivanova, 1968). Ten levels of categorisation can be used: class, sub-class, range, type, subtype, genus, species, variety, category and phase. However, type is the highest level currently used. Subtypes are defined in terms of intensity of humus and organic matter accumulation, nature and type of translocated carbonates, gypsum and soluble salts, and depth of winter cooling and summer heating.

The Soil Taxonomy (SCS)

This widely used classification was started in 1951 and has evolved in a series of approximations. Essentially, it is the seventh approximation (1960) that is now used, but modifications are continually being made. The basis of the classification is somewhat confusing. In many parts of the classification, soil-forming processes are important; it is therefore partly a genetic classification. But there is also a

Table 5.2 Classification of soils by Dokuchaev

Zones	Soil type
Class A: Normal or zonal soils	
I Boreal	Tundra (dark brown) soils
II Taiga	Light grey podzolised soils
III Forest–steppe	Grey and dark grey soils
IV Steppe	Chernozem
V Desert–steppe	Chestnut and brown soils
VI Aerial or desert zone	Aerial soils, yellow soils, white soils
VII Subtropical and zone of tropical soils	Laterite or red soils
Class B: Transitional soils	
VIII	Dryland moor soils or moor meadow soils
IX	Carbonate-containing soils (rendzina)
X	Secondary alkali soils
Class C: Abnormal soils	
XI	Moor soils
XII	Alluvial soils
XIII	Aeolian soils

stress on the recognition and definition of a specific number of diagnostic horizons. These horizons can be divided into surface horizons, or epipedons, and subsurface horizons. Some of the more important diagnostic epipedons are defined briefly in Box 5.3. Epipedons form at the surface but are not synonymous with A horizons. They include the upper part of the soil that is darkened by organic matter but they usually also include the upper eluvial horizons and may include part of the illuvial B horizon if there is enough darkening by organic matter. A number of subsurface diagnostic horizons are also used in the Soil Taxonomy scheme, some of which are defined in Box 5.4. Other horizons or layers often have diagnostic value, such as duripans and fragipans.

The Soil Taxonomy scheme is essentially a hierarchical classification, with soils divided into orders, suborders, great groups, subgroups, families and series. The way the division operates is shown in Figure 5.4 with respect to spodosols (podzols). There were originally ten orders, but lately an extra order, andisols, has been added (Table 5.3). The names of all the orders have a common ending in -sol, e.g. spodosol. For suborders, the formative element is taken from the name and used as an ending for suborders, great groups and subgroups. Thus for spodosols, it is the -od- that is used. In suborders, the first part of the name indicates the property of the class and the second is the formative element. Thus the suborder orthods (Gk *orthos* – true) are the common spodosols. Names of great groups are formed by the addition of one or more prefixes to the name of the suborder: e.g. cryorthods – cold orthods (Gk *kruos* – cold). The subgroup takes the name of the great group preceded by one or more adjectives: thus typic cryorthods – typical cryorthods.

Orders are differentiated by the presence or absence of diagnostic horizons or morphological features (e.g. histosol, mollisol, spodosol), features that show the dominant set of soil-forming processes that have taken place (inceptisol, vertisol, ultisol) or chemical properties (alfisol, oxisol). This is essentially a subjective process, as there are no fixed

Box 5.3

SURFACE DIAGNOSTIC HORIZONS (EPIPEDONS)

Mollic epipedon

A relatively thick, dark-coloured surface horizon. The dark colour is due to the presence of organic matter. The horizon is usually rich in base cations – calcium, magnesium and potassium – so base saturation is over 50 per cent. Structure is usually granular or blocky.

Umbric epipedon

A dark surface horizon resembling the mollic epipedon but with base saturation less than 50 per cent.

Histic epipedon

A thin surface horizon of peat. The horizon is saturated with water for 30 consecutive days or more during the year. Very thick accumulations of peat are regarded as organic soils and are not classified as epipedons.

Ochric epipedon

A surface horizon that is light in colour and contains less than 1 per cent organic matter. Surface horizons that are too thin, too dry or too hard to qualify as any one of the other listed epipedons are also included.

Plaggen epipedon

A human-made surface layer greater than 50 cm thick. It has been produced by long-continued manuring.

Box 5.4

SUBSURFACE DIAGNOSTIC HORIZONS

Argillic horizon

An illuvial horizon (usually the B horizon) in which clay minerals have accumulated by illuviation. Clay cutans (argillans) are usually present.

Argic horizon

An illuvial horizon formed under cultivation and containing illuvial silt, clay and humus. Ploughing helps the downwashing of these materials, which accumulate beneath the ploughed layer.

Natric horizon

A natric horizon is similar to an argillic horizon but has a prismatic structure and a high proportion of Na^+, amounting to 15 per cent or more of the CEC.

Calcic horizon

A horizon of accumulation of calcium carbonate or magnesium carbonate.

Petrocalcic horizon

A hardened calcic horizon that does not break up when soaked in water.

Gypsic horizon

A horizon of accumulation of hydrous calcium sulphate (gypsum).

Salic horizon

A horizon enriched by soluble salts.

Albic horizon

A pale, often sandy, horizon from which clay and free iron oxides have been removed. It is usually the E horizon.

Spodic horizon

A horizon containing precipitated amorphous materials composed of organic matter and sequioxides of aluminium with or without iron. It is formed partly by illuviation and usually underlies an E horizon.

Cambic horizon

An altered horizon with texture as fine as or finer than very fine sand that has lost sesquioxides or bases, including carbonates, by leaching. It is considered a B horizon, although it has accumulated little clay.

Oxic horizon

A highly weathered horizon at least 30 cm thick, rich in clays and sesquioxides of low CEC (16 or less). Few primary minerals are left to release bases. Oxic horizons are old and tend to be restricted to equatorial, tropical and subtropical zones.

principles. Orders are divided into forty-seven suborders. Suborders are differentiated using criteria that vary from order to order. The number of suborders ranges from two to seven per order, with most having four or five. In the differentiation of great groups, the whole assemblage of horizons is considered, together with a number of diagnostic properties. Great groups are subdivided into subgroups by the addition of adjectives to the great groups' names (see Figure 5.4). Further subdivision into families occurs on the basis of particle size, mineralogy and temperature regimes. The final level of subdivision, the series, is achieved on the basis of

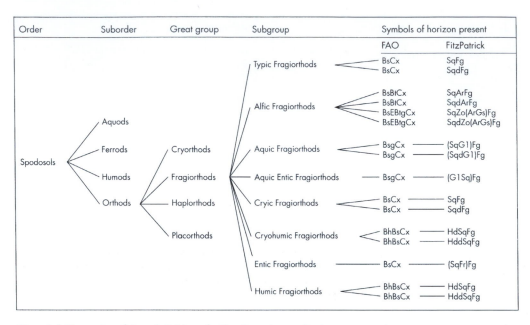

Figure 5.4 Illustration of the subdivision of soil orders using spodosols as an example.
Source: From FitzPatrick (1983)

Table 5.3 Soil Survey Staff (1975, 1992) soil orders, formative elements and connotations

Order	Formative element	Connotation
Alfisols	M. *Alf* (from Pedalfer)	Clay translocation
Andisols	M. *And* (from Andosol)	Volcanic soil
Aridisols	L. *aridus* (dry)	Arid soil
Entisols	M. *Ent* (from Recent)	Early stage of soil formation
Histosols	Gr. *histos* (tissue)	Limited organic matter decomposition
Inceptisols	L. *inceptum* (beginning)	Limited soil development
Mollisols	Gr. *mollis* (soft)	Dark brown/black surface horizon, soft when dry
Oxisols	Fr. *oxide* (oxide)	Oxic horizon or plinthite
Spodosols	Gr. *spodos* (wood ash)	Bleached eluvial horizon
Ultisols	L. *ultimus* (last)	Highly developed soil
Vertisols	L. *verto* (turn)	Turnover of soil

the locality in which that type of soil was first recognised. It has no real value in terms of soil classification or genesis but is used in mapping at more detailed scales. A brief summary of the characteristics of the soil orders is now presented. Global patterns are examined in Chapter 7.

Andisols

These form on parent material of volcanic origin, especially ash. They possess a fine texture, fresh weatherable minerals, low bulk densities, a significant organic matter content and a high cation exchange capacity. Such soils can be very fertile.

Alfisols

These soils generally possess an argillic horizon and moderate to high base saturation, and water is held at <15 bars tension for at least three months each year during the growing season. Fragipans and duripans may be present. The argillic horizon is enriched by accumulated clay minerals and is moderately saturated with exchangeable bases such as calcium and magnesium. Alfisols are common in humid temperate environments and are widespread in Western Europe, including eastern and southern Britain, eastern North America, central Siberia, northern China and southern Australia. They are also found in areas such as eastern Brazil; West, East and southern Africa; western Madagascar; northern Australia; eastern India and Southeast Asia associated with wet–dry tropical climates and semi-arid subtypes of tropical and subtropical dry climates. They are often highly fertile, with a stable soil structure, and they contain significant quantities of mull humus. In low-lying areas, especially if they possess an argillic B horizon, waterlogging may occur. There are five suborders: aqualfs, boralfs, xeralfs, udalfs and ustalfs. Boralfs are the alfisols of cold forest lands, with grey surface horizons and a brownish subsoil. Udalfs occur in mid-latitude zones formed originally under deciduous forest. These soils are now extensively cultivated. Ustalfs are brownish to reddish alfisols of warmer climates

and are usually associated with wet–dry tropical climates. Xeralfs are associated with a xeric soil water regime under mediterranean climates. They are usually brownish or reddish in colour.

Aridisols

These are the soils of dry areas, both cold and hot, where rainfall, for various reasons, hardly penetrates into the soil. They may have a high salt content, and for most of the year water is held at a tension of >15 bars. Profile morphology is extremely varied, but they usually possess a pale or anthropic upper horizon. Carbonate accumulations are often present at depth. They may also possess properties inherited from previous environmental conditions. Poor management results in soil degradation, salinisation and sodification. There are two suborders: argids and orthids. Argids possess an argillic horizon, where clay minerals have accumulated by illuviation and have often formed on surfaces of late Pleistocene or even older age. Illuviation is thought to have occurred during moist (pluvial) periods. Orthids do not possess this argillic horizon. One variety, salorthids, has a salic horizon of salt accumulation.

Entisols

Entisols show little or no evidence of the development of middle horizons. Most have no horizons due to their youth, steepness of slope, active erosion or perhaps floodplain deposition. They may also be found in cold environments, where soil formation is limited. For these reasons, they are found throughout the full global range. They have low water-holding capacity and low nutrient status. There are five suborders: aquents, arents, fluvents, orthents and psamments. Fluvents are formed on recent river alluvium, orthents are common on glacially abraded surfaces, and psamments develop on dune and beach sand. The Sand Hills region of Nebraska is a major area of psamments.

Histosols

These are composed predominantly of organic matter and are often saturated with water. They can form in any environment where organic matter additions to the soil are considerable, where organic decomposition is slow and where water is present. They are especially common in cool, moist areas such as northern Canada. The four suborders of fibrists, folists, hemists and saprists are defined on the basis of moisture regime and degree of decomposition of organic matter. Typical examples occur in upland bogs, moorland, low fen and carr. Some are highly acid and nutrient-poor, whereas some are nutrient-rich. Draining such soils leads to rapid decomposition (oxidation) of the organic matter and shrinkage (see Chapter 8).

Inceptisols

These are soils of humid regions with altered horizons that have lost material by leaching but which still contain some weatherable minerals. The clay fraction has a moderate to high CEC. Soils in this order are extremely variable and are found over a wide range of latitudes, usually on very young geomorphological surfaces. Morphological expression of soil development is poor. They are widespread on the floodplains of major rivers such as the Ganges, Mississippi and Amazon and the deltaic plains of the Nile, Irrawaddy and Mekong. There are six suborders: andepts, aquepts, ochrepts, plaggepts, tropepts and umbrepts. Tropepts are brownish to reddish, freely drained and thin, and they occur on steep slopes of Pleistocene and Holocene age, especially in hilly tropical areas. Umbrepts are acid, dark reddish or brownish, freely drained and rich in organic matter, and they occur mainly in hilly to mountainous regions of humid mid- to high altitudes on late Pleistocene or Holocene deposits under coniferous forest. Aquepts form in wet places. Cryaquepts, aquepts of cold environments, are sometimes called tundra soils.

Mollisols

These are the dark-coloured, base-rich soils of temperate grasslands. They also occur at high latitudes and high altitudes, and in intertropical areas. They are rich in mull humus with a well-developed crumb structure in the upper horizons. They usually contain a high level of calcium and large earthworm populations. Classic chernozems are mollisols. They can be regularly exposed to drought and those developed on loess are susceptible to wind erosion. There are seven suborders: albolls, aquolls, borolls, rendolls, udolls, ustolls and xerolls. Borolls are the mollisols of the cold-winter semi-arid plains and steppes. Ustolls contain calcium carbonate, starting at about 50 cm, which may form a petrocalcic horizon sometimes known as caliche. Xerolls form in inland areas under a mediterranean climate, such as northern parts of California and Nevada and southern Idaho. In North America, mollisols dominate the Great Plains, Columbia Plateau and northern Great Basin. In South America, they cover large areas of the pampas of Argentina and Uruguay. In Eurasia, they correspond to the steppes of Russia, Siberia and Mongolia.

Oxisols

These are red, yellow or grey soils of moist tropical and subtropical areas. They generally occur on gently sloping areas of great age, are strongly weathered, contain quartz, kaolinite and oxides of aluminium and iron and are often acidic, with a low nutrient status. The CEC of the clay fraction is extremely low. They usually possess an oxic horizon within 2 m of the surface and often a plinthite within 30 cm. Because they occur on surfaces of great age, they have experienced a number of climate changes. For an oxisol to develop the climate must be moist, but they also occur in seasonally dry environments because a climate change has occurred since their time of formation. There are five suborders: aquox, humox, orthox, torrox and ustox.

Plate 1 A red latosol, Tanzania
(M.A. Oliver)

Plate 2 Yellow podzol with a distinct plough layer and organic staining, Tanzania *(M.A. Oliver)*

Plate 3 A podzol from the New Forest, England, showing a prominent bleached eluvial layer and iron pan
(M.A. Oliver)

Plate 4 An earthy sulphuric peat soil with mineral layers. USDA, Typic Sulfohemist; FAO, Dystric Histosol
(M.A. Oliver)

Plate 5 Extremely thin soil developed on rubbly chalk
(M.A. Oliver)

Plate 6 Typical brown earth; fine loamy material over lithoskeletal siltstones and sandstones, Wyre Forest, England. USDA, Typic Dystrochrept; FAO, Dystric Cambisol
(M.A. Oliver)

Plate 7 Rendzina developed on limestone, Corfu
(M.A. Oliver)

Plate 8 Blanket peat, Scotland, showing differing degrees of humification
(M.A. Oliver)

Plate 9 Blocky structure with pronounced cracking in a red plateau soil, Tanzania
(M.A. Oliver)

Plate 10 Termite mound, Tanzania
(M.A. Oliver)

Plate 11 Ice wedge cast, Thetford Forest, eastern England
(M.A. Oliver)

Plate 12 Well-developed stone line in red plateau soil, Tanzania
(M.A. Oliver)

Plate 13 Frost hummock, Iceland, showing cryoturbated volcanic ash and windblown material
(J. Gerrard)

Plate 14 Wind-eroded exposure, southern Iceland, showing prominent white and black volcanic ash layers and aeolian deposits
(J. Gerrard)

Plate 15 Well-developed granular structure in topsoil
(M.A. Oliver)

Plate 16 Blocky structure in topsoil
(M.A. Oliver)

Spodosols

Soils in this order often possess a peaty surface horizon with mor humus and poorly decomposed plant remains. They are podzols in the old terminology. They generally possess a bleached grey to white eluvial (albic) horizon and are coarse-textured, highly leached and acidic, with a low nutrient status. They are characterised by a spodic B horizon of accumulation of organic material and compounds of aluminium and iron. They form in the cool temperate boreal forests of North America and Eurasia. As they occur in regions that have recently experienced glaciation, they are comparatively young. There are four suborders: aquods, ferrods, humods and orthods.

Ultisols

Such soils represent the ultimate or end stage of weathering. They contain few weatherable minerals, thus they are highly weathered, leached and acid, with a low nutrient status, and they are often dominated by the clay mineral kaolinite. They usually possess an argillic horizon. The B horizons of well-drained ultisols are red or yellowish brown in colour due to concentrations of oxides of iron. They have developed over a long period in humid warm temperate to tropical environments. Extensive areas of these soils occur in the southeastern USA, eastern and southern Brazil, India and parts of Southeast Asia, and northeastern Australia. There are five suborders: aquults, humults, udults, ustults and xerults.

Vertisols

These are clayey (swelling) soils possessing deep, wide cracks. The swelling clays (usually smectite) expand and contract on wetting and drying, respectively, turning over the soil and producing deep cracks with evidence of movement such as slickensides and structural aggregates tilted at an angle. In the latest definition (Soil Survey Staff, 1994), a linear extensibility (the product of thickness of soil layer multiplied by the coefficient of linear extensibility (COLE, see Chapter 3) summed for all soil horizons) of 6 cm or more is offered as an alternative to the usual morphological requirements of cracks, slickensides and wedge-shaped aggregates. The soil between cracks has a high bulk density. A distinctive pattern, known as gilgai, is also produced. Gilgai are small surface features that may be hummocks, narrow ridges or basins. They are seldom higher than 1 m. Vertisols are low in organic matter but with a high base saturation. They occur in warm temperate to tropical environments, including arid areas, and are especially common in parts of Australia, India, Sudan and the southern coastal regions of the USA. In India, they are especially common on the weathered basalts of the Deccan region. They are called by a variety of other names, such as black cotton soils, tropical black earths and regur. There are four suborders: torrerts, uderts, usterts and xererts.

FAO/UNESCO

The most recent FAO/UNESCO (1989) scheme is a very widely used system. It is based on an earlier scheme (FAO/UNESCO, 1974) and was designed primarily to be used for the production of the *Soil Map of the World*. This was created to show units that were sufficiently broad to have general validity and contain sufficient elements to reflect as closely as possible the soil pattern of a large region. It employs twenty-eight major soil groupings, which are subdivided into 153 units. It can be seen (Table 5.4) that the names of the major groupings are derived from a number of linguistic roots. Some have been used before in other schemes; others have been newly devised. Approximate equivalents with the Soil Taxonomy orders are shown in Table 5.5.

An attempt was made to use as many 'traditional' names as possible, such as chernozems, podzols, planosols, solonetz, solonchaks, rendzinas, regosols and lithosols. Names such as vertisols, rankers and ferralsols, which have gained recent recognition, have also been kept. Thus those familiar with the traditional soil names may find this scheme more

Table 5.4 Major FAO/UNESCO soil groupings, formative elements and connotations

Major soil grouping	Formative element	Connotation
Acrisols	L. *acer, acetum* (strong acid)	Low base saturation
Alisols	L. *alumen*	High aluminium content
Andosols	Jap. *an* (dark), *do* (soil)	Dark surface horizon (rich in volcanic glass)
Anthrosols	Gr. *anthropos* (man)	Resulting from human activities
Arenosols	Gr. *arena* (sand)	Weakly developed, coarse-textured soils
Calcisols	L. *calx* (lime)	Calcium carbonate accumulation
Cambisols	L. *cambiare* (to change)	Changes in colour, structure and consistency
Chernozems	Russ. *chern* (black), *zemlja* (earth land)	Black, rich in organic matter
Ferralsols	L. *ferrum, alumen*	High sesquioxide content
Fluvisols	L. *fluvius* (river)	Alluvial deposits
Gleysols	Russ. *gley* (mucky soil mass)	Excess water
Greyzems	AS *grey*; Russ. *zemlja* (earth, land)	Uncoated silt and quartz within organic-rich layers
Gypsisols	L. *gypsum*	Calcium sulphate accumulation
Histosols	Gr. *histos* (tissue)	Fresh or partly decomposed organic matter
Kastanozems	L. *castanea* (chestnut); Russ. *zemla* (earth, land)	Organic-rich, brown colour
Leptosols	Gr. *leptos* (thin)	Weakly developed, shallow soils
Lixisols	L. *lixivia* (washing)	Clay accumulation and strong weathering
Luvisols	L. *luvere* (to wash)	Clay accumulation
Nitisols	L. *nitidus* (shiny)	Shiny ped faces
Phaeozems	Gr. *phaios* (dusky), Russ. *zemlja* (earth, land)	Organic-rich, dark colour
Planosols	L. *planus* (flat, level)	Seasonal surface waterlogging on level or depressed relief
Plinthosols	Gr. *plinthos*	Mottled clayey materials harden on exposure
Podzols	Russ. *pod* (under), *zola* (ash)	Strongly bleached horizon
Podzoluvisols	Podzols and luvisols	
Regosols	Gr. *rhegos* (blanket)	Loose mantle of material
Solonchaks	Russ. *sol* (salt)	Salty area
Solonetz	Russ. *sol* (salt), *etz* (strongly)	
Vertisols	L. *vertere* (to turn)	Turnover of surface soil

Table 5.5 Approximate relationships between the Soil Survey Staff (1975, 1992) soil orders and FAO/UNESCO (1989) major soil groupings

Soil Survey Staff soil orders	FAO/UNESCO major soil groupings
Alfisols	Luvisols
Andisols	Andosols
Aridisols	Calcisols, gypsisols, solonchaks, solonetz
Entisols	Arenosols, fluvisols, leptosols, regosols
Histosols	Histosols
Inceptisols	Cambisols
Mollisols	Chernozems, greyzems, kastanozems, phaeozems
Oxisols	Alisols, ferralsols, nitosols, plinthosols
Spodosols	Podzols
Ultisols	Acrisols, lixisols
Vertisols	Vertisols

user-friendly. Other soil terms, such as brown forest, podzolic, prairie, mediterranean, lateritic and alluvial, were not retained, because of confusion generated by their dissimilar use in different countries. To take two examples: 'brown forest soils', as a term, has been used to describe a wide variety of different soils. Initially it was used to describe soils 'developing in subhumid temperate climates, having a "mull" humus, a B horizon with a strong coloration and a slightly higher clay content than the C horizon, but showing no signs of clay illuviation, and having calcium carbonate in the lower part of the solum' (FAO/UNESCO, 1974: p.11). But it has also been used to describe acid brown forest soils, tropical soils and podzolised brown forest soils. Similarly, the term 'podzolic' has been used to indicate illuvial clay accumulation, the formation of a bleached horizon, the penetration of bleached tongues of eluvial material into a B horizon, an abrupt textural change between the eluvial and B horizon, and illuviation of acid organic matter and oxides and hydroxides of iron and aluminium.

Many of the characteristics used to define the soils are morphological, such as those based on texture, structure or colour (arenosols, cambisols). Others refer to processes (luvisols, vertisols) or sometimes chemistry (acrisols, ferralsols). The division of soil groups into units is achieved on the basis of quantitatively defined diagnostic horizons and properties. Five types of surface horizon, five types of B horizon and six other types are used, as well as twenty-six diagnostic properties based on morphological and chemical properties. The diagnostic horizons used are essentially the same as those for Soil Taxonomy. This scheme is less hierarchical than many other schemes and seems to work well at the scale adopted in the *Soil Map of the World* (1:1,000,000). It is the scheme adopted for the discussion of world soils in the next chapter.

Other traditional schemes

Most of the schemes so far discussed are based on a hierarchy of groupings. A number of other national hierarchical schemes have been devised, such as South Africa (Soil Classification Working Group, 1991), New Zealand (Hewitt, 1992), France (CPCS, 1967; Duchaufour, 1982), the Netherlands (de Bakker and Schelling, 1966) and Germany (Mückenhausen, 1985). Three other important schemes, which are now examined, are those for the British

Isles (Avery, 1990), Canada (Canada Soil Survey Committee, 1978) and Australia (Isbell *et al.*, 1997).

The current classification of British soils (Avery, 1990) is based on the former England and Wales system (Avery, 1980). The major soil groups and groups are listed in Table 5.6, with US Soil Taxonomy and FAO/UNESCO equivalents. Groups are further subdivided into subgroups.

Lithomorphic soils comprise almost all of the soils that have been grouped as lithomorphic soils in the former England and Wales system (Avery, 1973; 1980), as immature soils and rendzinas by the Soil Survey of Scotland (1984) and as lithosols, regosols and rendzinas in Ireland (Gardiner and Radford, 1980). Their common feature is the presence of bedrock or little altered regolith at shallow depth. They are restricted to young surfaces affected by recent erosion or deposition and to other places where pedogenesis has been restricted by the nature of the bedrock, such as on chalk (rendzinas) and hard siliceous rocks (lithosols and rankers). Associations consisting predominantly of lithomorphic soils, including bare rock, unvegetated screes and raw coastal sands, cover about 7 per cent of England and Wales. Some lithomorphic soils are uncultivable, and for the others cultivation is restricted by shallowness, stoniness or rockiness, small available water capacity, and/or liability to erosion or flooding. There are five groups.

Brown soil refers to more or less well-drained soils with altered subsurface horizons, usually brown or reddish, in which iron oxides are bonded to silicate clays. Some soils have a colour/structure B horizon (Bw) and others an argillic (Bt) or podzolic B (Bs) horizon. A few lack a distinct B horizon but possess a prominent cumulic A (or A/B) horizon. The major group includes calcareous soils as well as the originally non-calcareous or decalcified soils traditionally grouped in Britain as brown earths. Brown soils cover approximately 45 per cent of England and Wales, about 40 per cent of Ireland and between 10 and 15 per cent of Scotland. They occur mainly in lowland areas with subhumid or humid temperate climates. There are six soil groups.

Podzols, as classified here, are required to possess a podzolic B horizon (Bh and/or Bs), a thin iron pan (Bf), or both. If the B horizon consists only of an uncemented (friable) Bs, there must be an overlying albic E (Ea or Eag) at least 5 cm thick. At the group level, they are divided into non-hydromorphic podzols, gley podzols and stagnopodzols. The first group are pervious and well aerated; gley podzols possess a gleyed horizon below a podzolic B, and stagnopodzols possess characteristics indicating anaerobic conditions above a thin iron pan but with few or no signs of gleying at depth. There are ten subgroups, differentiated on the basis of the kind and sequence of subsurface horizons. Podzols cover approximately 5 per cent of England and Wales, 25 per cent of Scotland and about 8 per cent of Ireland. In the subhumid lowlands, they are restricted to coarse-textured siliceous sediments and finer, base-deficient deposits where the clay content of the uppermost horizons has been reduced by previous eluviation. In cooler and more humid areas they occur on a wider range of materials but usually in well-drained sites with pervious substrata.

Gley soils are classified as such by the presence of a gleyed or hydrocalcic subsurface horizon that commences directly below the topsoil or within 40 cm depth. They are periodically or permanently saturated with water or are formed under wet conditions but lack horizons characteristic of podzols (Avery, 1990). Gley soils have been recognised using similar criteria in all three soil survey organisations in the British Isles. Gley soils cover about 40 per cent of England and Wales and approximately 30 per cent of Scotland and Ireland. Their distribution is governed by current or former soil-water regimes, by subsoil permeability and by rainfall/ evapotranspiration relationships. The soils can be divided into surface water (pseudogley, stagnogley) and groundwater (gley *sensu stricto*) types. There are five soil groups.

Man-made soils, as a classification, are restricted to predominantly mineral soils 'with distinctive features that can be attributed to the recent or former incidence of "abnormal" land-use or management practices such as addition of earth-containing

Table 5.6 Classification of soils according to the scheme adopted in the British Isles and US Soil Taxonomy and FAO/UNESCO equivalents

Major soil group	Soil group	US Soil Taxonomy	FAO/UNESCO
Lithomorphic soils	Lithosols	Very shallow (lithic) orthents	Lithosols
	Rankers	Mainly shallow (lithic) umbrepts and orthents; some lithic histosols and hapludolls	Mainly rankers and regosols (dystric or euric); some lithic histosols and phaeozems
	Rendzinas	Rendolls and calcareous orthents; some udifluvents	Rendzinas and calcaric regosols; some fluvisols
	Sandy regosols	Mainly psamments; some orthents	Sandy regosols
	Rego-alluvial soils	Fluvents, orthents or psamments	Fluvisols or regosols in recent alluvial deposits
Brown soils	Sandy brown soils	Mainly psamments; some orthents and sandy umbrepts and hapludolls	Mainly arenosols; some sandy phaeozems and humic cambisols
	Alluvial brown soils	Mainly fluventic ochrepts; some udifluvents and hapludolls	Mainly cambisols; some fluvisols and phaeozems
	Calcaric brown soils	Mainly eutochrepts; some hapludolls and udifluvents	Mainly calcic cambisols; some phaeozems and fluvisols
	Orthic brown soils	Mainly non-calcareous ochrepts; some umbrepts, udifluvents and hapludolls	Mainly cambisols; some phaeozems and fluvisols
	Luvic brown soils	Mainly udalfs; some udults and argiudolls	Luvisols, acrisols and podzoluvisols
	Podzolic brown soils	Mainly dystochrepts or fragio-chrepts; some orthods and andepts	Mainly dystric cambisols; some leptic podzols and andosols
Podzols	Non-hydromorphic soils	Humods and orthods, excluding placic subgroups	Orthic and humic podzols
	Gley podzols	Mainly haplaquods and sideraquods	Gleyic podzols

continued

Table 5.6 continued

Major soil group	Soil group	US Soil Taxonomy	FAO/UNESCO
Podzols (continued)	Stagnopodzols	Placaquods and placaquepts; some placic humods, sideraquods and humaquepts	Placic podzols and some gleyic podzols and humic gleysols
Gley soils	Sandy gley soils	Psammaquents, sandy haplaquolls and humaquepts; some haplaquents	Sandy gleysols
	Alluvial gley soils	Fluvaquents, hydraquents, fluventic haplaquolls and humaquepts	Fluvisols and mollic or humic gleysols in recent alluvial deposits
	Calcaric gley soils	Calcareous haplaquepts and hapladolls	Calcaric and mollic gleysols
	Orthic gley soils	Mainly non-calcareous aquepts; some haplaquolls	Non-calcareous gleysols
	Luvic gley soils	Mainly aqualfs and aquepts; some aqaults and argiaquolls	Mainly gleyic luvisols or acrisols and planosols
Man-made soils	Cultosols	Mainly plaggepts	no related unit
	Disturbed soils	Mainly arents	no related unit
Peat soils	Fen soils	Histosols, mainly euic or sulfic	Eutric and some dystric histosols
	Bog soils	Histosols, mainly dysic	Dystric histosols

Source: After Avery (1990)

manures, unusually deep cultivation, or wholesale removal and re-emplacement of soil material' (*ibid.*: p. 385). They possess thick man-made A horizons, an artificially reworked layer, or both. They are subdivided into cultosols, equivalent to the man-made humus soils in the England and Wales classification, and disturbed soils.

Peat soils are organic soils formed from plant remains accumulated under wet conditions, either as autochthonous peat in the position of growth or as part of sedimentary deposits such as lake muds. As defined in this classification, they possess at least 40 cm of organic material within the upper 80 cm, or at least 30 cm if it rests directly on bedrock, and no overlying mineral layer that is more than 30 cm thick and has a non-humose B or C horizon at its base. They are divided into two groups. Soils so defined cover about 3 per cent of England and Wales (Mackney *et al.*, 1983). They are more extensive in Scotland and Ireland, but areas cannot be estimated because slightly different criteria are used by the systems in Scotland and Ireland. Using the criteria of those systems, peat soils occupy about 10 per cent of Scotland and 16 per cent of Ireland (Hammond, 1981).

The aim of the Canadian system is to organise soils in a reasonable and usable way. Canada lies entirely north of the 40th parallel, therefore there is no need to use soil orders that occur only in lower latitudes. Also, the large expanse of Canada lying within boreal forest and tundra climates has resulted in a greater emphasis on soils characteristic of those regions. As a low percentage of the area is cultivated, the Canadian system does not recognise as diagnostic those horizons strongly affected by human activity such as ploughing.

Nine soil orders make up the highest taxon, divided into twenty-eight great groups (Figure 5.5). Brunisolic soils are well to imperfectly drained and occur in a wide range of climatic and vegetation environments, including boreal forest, mixed forest, shrubs and grass, as well as heath and tundra. They show a weak B horizon of accumulation and lack the diagnostic podzolic B horizon. Chernozemic soils are the classic chernozems of grasslands of the interior plains of western Canada. They possess a good accumulation of organic matter in surface horizons, producing dark colours. The A horizon is at least 10 cm thick. Calcareous accumulations occur in the C horizon. Soils of the cryosolic order occur over

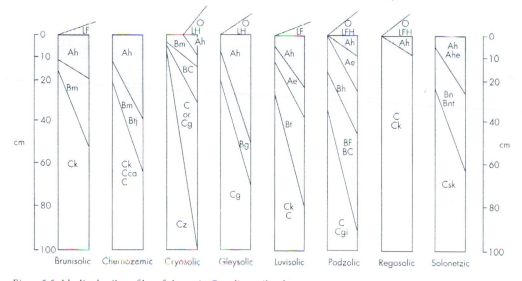

Figure 5.5 Idealised soil profiles of the main Canadian soil orders.

much of the northern third of Canada, where permafrost is close to the surface. Cryoturbation is common and patterned ground frequent. They may be mineral or organic soils, and organic matter decomposition will be slow. Gleysolic soils have characteristics indicating periodic or prolonged water saturation. They tend not to dominate large areas but occur as patches within areas of other soils. Luvisolic soils possess light-coloured eluvial horizons and illuvial B horizons where clay has accumulated. They develop in base-saturated sandy loam to clay parent material under forest vegetation in a variety of climates from humid to subhumid and from mild to very cold. The largest area of these soils occurs in the central and northern interior plains under deciduous, mixed and coniferous forest. Soils of the organic order are composed largely of organic materials (> 30 per cent organic matter by weight). They are peat and bog soils and are found in poorly drained depressions. Podzolic soils are the 'classic' podzols with B horizons enriched in organic matter, iron and aluminium and with an eluviated upper horizon. They tend to occur under forest and heath vegetation in cool to very cold humid climates. Regosolic soils are poorly developed soils lacking in genetic horizons. This may be because soil processes have been inhibited by the instability and youthfulness of parent materials, by the nature of parent material such as pure quartz sand or by dry, cold climatic conditions. Such soils have typical AC profiles with a surface organic horizon. They correspond to entisols of the Soil Taxonomy. Soils of the solonetzic order are the same as solonetz soils in other schemes, occurring on saline parent material in some areas of the semi-arid and subhumid interior plains.

A number of classification schemes have been used in Australia, and these have been reviewed by Isbell (1992). At the time of that review, two classification schemes were in wide use. The *Handbook of Australian Soils* (Stace *et al.*, 1968) was essentially a revision of the earlier great soil group scheme (Stephens, 1953). The *Factual Key* (Northcote, 1979) dates from 1960 and was based on a set of about 500 profiles, mostly from southeastern

Australia. The advantages and disadvantages of the two schemes have been discussed by Moore *et al.* (1983).

Since these classifications were developed, much more information has become available and has been incorporated into a new national soil classification (Isbell *et al.*, 1997). It is a multi-category scheme with classes defined on the basis of diagnostic horizons or materials and their arrangement in vertical sequence as seen in an exposed soil profile. Classes are based on real soil bodies, they are mutually exclusive, and the allocation of 'new' or unknown individuals to classes is by means of a key. All terms used in the classification are consistent with those defined in the second edition of the *Australian Soil and Land Survey Field Handbook* (McDonald *et al.*, 1990). It is a general-purpose hierarchical scheme with orders, suborders, great groups, subgroups and families. The name of the orders, the derivation of the name and connotation are shown in Table 5.7.

The B horizon is one of the most important features used in the classification. If B horizon material occupies more than 50 per cent (visual abundance estimate) of the horizon, that is it is a B, BC or B/C horizon, the soil is deemed to possess a B horizon and is classified accordingly. If the soil has a C/B horizon in which the B horizon component is between 10 and 50 per cent, the soil will be classified as a tenosol. If there is less than 10 per cent of B horizon material and no pedological development other than a minimal A1 horizon, the soil will be classified as a rudosol. The main orders are now examined briefly.

Anthroposols are soils resulting from human activities that have led to a profound modification, truncation or burial of the original soil horizons, or the creation of new soil parent materials by a variety of mechanical means. There are seven suborders. Calcarosols, as the name suggests, are soils that are usually calcareous throughout the profile, sometimes highly so. They are one of the most widespread and important soil groups in southern Australia. They are defined as soils that are calcareous throughout the solum, or at least directly

Table 5.7 Soil order nomenclature in the Australian soil classification scheme

Name of order	Derivation	Connotation
Anthroposols	Gr. *anthropos*, man	'man-made soils'
Calcarosols	L. *calcis*, lime	calcareous throughout
Chromosols	Gr. *chroma*, colour	often brightly coloured
Dermosols	L. *dermis*, skin	often with clay skins on ped surfaces
Ferrosols	L. *ferrum*, iron	high iron content
Hydrosols	Gr. *hydor*, water	wet soils
Kandosols	Kandite (1:1) clay minerals	–
Kurosols	–	pertaining to clay increase
Organosols	–	predominantly organic materials
Podosols	Russ. *pod*, under; *zola*, ash	podzols
Rudosols	L. *rudimentum*, a beginning	rudimentary soil development
Sodosols	–	influenced by sodium
Tenosols	L. *tenuis*, weak, slight	weak soil development
Vertosols	L. *vertere*, to turn	shrink–swell clays

Source: From Isbell *et al.* (1997)

below the A1 or Ap horizon, or a depth of 20 cm if the A1 horizon is only weakly developed. Carbonate accumulations must be pedogenic, and the soils do not possess clear or abrupt textural B horizons. There are seven suborders.

Chromosols possess a strong textural contrast between the A and B horizons. The B horizons are not strongly acid and are not sodic. These soils are widely used for agriculture, particularly those with red subsoils. They are defined as soils, other than hydrosols, with a clear or abrupt textural B horizon and in which the major part of the upper 20 cm of the B2 horizon (or the major part of the entire B2 horizon if it is less than 20 cm thick) is not sodic and not strongly acid. Soils with strongly subplastic upper B2 horizons are also included, even if they are sodic. There are five suborders, classified on the basis of colour. Dermosols possess structured B2 horizons that lack a strong textural contrast between the A and B horizons. It groups a wide range of soils with some important common properties. They are defined as soils, other than vertosols, hydrosols,

calcarosols or ferrosols, that have B2 horizons with structure more developed than weak throughout the major part of the horizon and clear or abrupt textural B horizons. There are five suborders, classified according to main colour.

Ferrosols have B2 horizons that are high in free iron oxide and lack a strong textural contrast between the A and B horizons. They are defined as soils, other than vertosols, hydrosols or calcarosols, that have B2 horizons in which the major part has a free iron oxide content greater than 5 per cent Fe in the fine earth fraction (< 2 mm) and do not have clear or abrupt textural B horizons or a B2 horizon in which at least 30 cm has vertic properties. There are five suborders, based on colour. Hydrosols are seasonally or permanently wet soils. The greater part of the profile is saturated for at least two to three months in most years. They may or may not experience reducing conditions for all or part of the period of saturation, and thus characteristics of reduction and oxidation conditions, such as gley colours and ochrous mottles, may or may not be present. There are seven suborders, classified mainly on the manner and nature of the water producing the saturated conditions.

Kandosols include soils that lack a strong textural contrast, have massive or weakly structured B horizons and are not calcareous throughout. Such soils are found throughout the Australian continent and often occur over large areas. They are defined as having B2 horizons in which the major part is massive or has only a weak grade of structure, possessing a maximum clay content in some part of the B2 horizon that exceeds 15 per cent, and they do not have a tenic B horizon. They are not calcareous, and there are five suborders. The kurosol order contains soils with a strong textural contrast between the A horizon and a strongly acid B horizon. Many kurosols have some unusual subsoil chemical features, such as high magnesium, sodium or aluminium contents. There are five suborders.

Organosols include most soils dominated by organic materials. They occur in small areas from the wet tropics to alpine regions but cover larger areas in southwest Tasmania. They are defined as having more than 40 cm of organic materials in the upper 80 cm or have organic materials extending from the surface to a minimum of 10 cm. In the latter case, the organic material usually directly overlies rock, partially weathered rock or saprolite, or other fragmental material. There are three suborders. Podosols possess B horizons dominated by the accumulation of compounds of organic matter, aluminium and/or iron. These are the 'podzols' found throughout the world. Australia is particularly noted for its 'giant' forms. There are three suborders, based on soil and site drainage conditions.

Rudosols accommodate soils with little pedological organisation. They are usually young soils, such that soil processes have had little time to modify the parent rocks or sediments. By definition, these soils grade to tenosols, and it may be a matter of judgement as to which order a particular soil is best placed in. There are nine suborders. Sodosols include soils with a strong textural contrast between the A horizon and sodic B horizon, which is not strongly

acid. Australia has many, widely distributed sodic soils, and the use of sodicity in Australian soil classification systems has been reviewed by Isbell (1995). There are five suborders.

The tenosol order has been devised to contain soils with generally only a weak pedological organisation apart from the A horizon. It accommodates a diverse but widespread range of soils. Tenosols differ from rudosols by having either a more than weakly developed A1 horizon, an A2 horizon or a weakly developed B horizon. They grade to kandosols, and it may be difficult to separate medium-textured tenosols from kandosols. There are six suborders. Vertosols embrace clay soils with shrink–swell properties exhibiting strong cracking when dry and at depth have slickensides and/or lenticular structural aggregates. Australia possesses the greatest area and greatest diversity of cracking clay soils of any country of the world. There are six suborders.

Alternative schemes

A rather different scheme was devised by FitzPatrick (1971) to try to counter some of the problems inherent in most of the schemes so far examined. It has yet to receive wide endorsement, but the ideas that it puts forward are worth considering because they say something fundamental about the nature of soils and the way they develop. The scheme is based on the generally held belief that soils form a continuum over the surface of the Earth and that what is required is some sort of basic descriptive scheme without any preconceived notion of soils fitting into some philosophical generic model. Soil can be considered as being composed of infinite coordinates in hyperspace. As Crowther (1953) has stressed, soil properties can be regarded as coordinates with arbitrary divisions creating spaces or segments with defined values. With respect to tropical soils, four separate coordinates might be colour, clay content, CEC and weatherable minerals.

Interestingly, the scheme independently devised by FitzPatrick has many similarities to one devised by the Russian worker Sokolovsky (1930). As has been seen, most major classification schemes use the

ABC horizon system together with diagnostic horizons. But, as noted, there are problems with such a scheme, and many workers have suggested its abandonment or modification. Sokolowsky suggested that horizons should be recognised and designated by the first letter of each horizon; e.g. H = humus, E = eluvial horizon, I = illuvial horizon, Gl = gley horizon, G = gypsum, P = parent rock, etc. Such letters would be used in combination with subscripts indicating the thickness of horizons in centimetres. Thus a podzol might be designated $H_{14}E_{14}E1_{40}I_{40}IP$.

In FitzPatrick's scheme, the basic theoretical unit becomes a segment that is 'a part of the conceptual model of soils having defined ranges of properties created on coordinate principles and embracing a number of individual horizons.' There are two basic types of segment. A reference segment is one that has a single unique dominating property or a unique combination of dominating properties and is formed principally by a single set of processes. An intergrade segment contains properties that grade between two reference segments. Four types of horizon, reference, intergrade, compound and composite, are used. A reference horizon is one that has properties that fall within a reference segment. Similarly, an intergrade horizon is one that has properties that fall within an intergrade segment. A compound horizon contains a combination of properties of two or more reference segments. Properties are often contrasting and develop possibly as a result of contrasting seasonal processes, or one set of properties may be superimposed upon an earlier

set. A composite horizon is one that contains discrete volumes of two or more segments. Intergrade horizons are designated by combining the appropriate symbols in parentheses according to the proportion of each set of properties, the predominant set being placed first. For compound horizons, the symbols are put in curly brackets and for composite horizons the symbols for the discrete volumes of each horizon are placed in round brackets and separated by a forward stroke.

The names of reference horizons are based on some conspicuous or unique property, but all names end with '-on': thus mullon, fermenton and humifon for various humic horizons. Taking the example of a podzol, it might be designated as $Lt_2Fm_3Hf_2$ $Mo_7Zo_5Sq_{40}$-As (Lt = litter; Fm = fermenton; Hf = humifon; Mo = modon (FAO umbric A_1 horizon; USDA umbric epipedon); Zo = zolon (FAO albic E horizon; USDA albic horizon); Sq = sesquon (FAO spodic B horizon, USDA spodic horizon)). Each of these horizons has its own characteristics. A horizon key has been provided for use in the field. Texture patterns can also be included without reference to the processes involved, which may be unknown (Box 5.5). The < and > symbols are placed between the horizon symbols and used singly or in pairs.

As FitzPatrick (1983) has stressed, the divisions do not delimit discrete entities, as the situation defies classification. The best that can be achieved is to provide each soil with a designation and to provide one or two higher levels of grouping using *ad hoc* methods. This places greater emphasis on the soil itself. In many ways, treating soil as a

---

Box 5.5

REPRESENTATION OF TEXTURE PATTERN IN THE SCHEME DEVISED BY FITZPATRICK

<   the horizon above has slightly less clay or coarser texture than the horizon below.
<<   the horizon above has much less clay or a much coarser texture than the horizon below.
>   the horizon above has slightly more clay of a finer texture than the horizon below.
>>   the horizon above has much more clay or a much finer texture than the horizon below.

continuum in hyperspace was a precursor to more recent attempts at classifying soils using numerical methods (see below).

Problems with these classification schemes

There are major problems with many, if not all, of the main classification schemes so far produced. Most of these have been summarised by FitzPatrick (1983) and Ellis and Mellor (1995). The most serious can be listed as follows:

1   Most employ a difficult and cumbersome terminology, which must militate against their efficient use.
2   The critera used to distinguish soils at one level in a hierarchy are not defined using the same characteristics, e.g. in the FAO/UNESCO scheme a number of different characteristics are used to define the major groups.
3   Many differentiating criteria are a mixture of easily identifiable properties and those requiring detailed analysis. These are not always easy to use, certainly not in a field situation.
4   The use of quantitative criteria suggests that significant thresholds exist in the values of certain properties and in their significance to soil formation. But, as has been stressed, soil forms a continuum. Thus soils that are very similar may be assigned to different classes because one soil just fails to meet a specific quantitative criterion. A specific example noted by many workers (e.g. Avery et al., 1977; McKeague et al., 1983; Lietzke and McGuire, 1987) is where podzols cannot be identified in the Soil Taxonomy spodosol category because B horizons do not meet the criteria used in defining a spodic horizon. There is also the reverse possibility of different soils being placed in the same category.
5   Complexity varies between the subdivisions of major soil units.
6   The paucity of information supplied by the schemes.
7   Inadequate provision for intergrades in spite of

their importance. Thus soils will be pushed into categories that are not appropriate.
8   Most schemes are hierarchical and soils are not.

It is for many of these reasons that attempts have been made to produce numerical soil classifications.

Numerical classifications

Most numerical classifications seek to produce classes within which the members are generally alike and substantially different from the members of other classes. Because of this, these methods are often known by the general name of cluster analysis. Clusters can be produced by either a hierarchical or a non-hierarchical method. As seen previously, most soil classifications have adopted a hierarchical method, and numerical methods to create similar structures have been popular. But hierarchical methods may not be the most appropriate or efficient. Here, the two methods are described and then evaluated.

Hierarchical methods

A hierarchical classification is one in which individuals belong to small groups, the small groups belong to larger groups and so on (Webster and Oliver, 1990). Grouping is usually made at a few distinct levels of generalisation, known as categories. Hierarchies can be produced by agglomerative methods, grouping individuals together into larger and larger groups, or by division to produce smaller and smaller groups from a single population. In general, agglomerative grouping is more efficient and is more widely adopted. With divisive grouping, there is the risk that important clusters will be split. This is the main reason why divisive methods have not been used for classifying soils.

Before a classification can be produced, the large number of measurements often taken within a particular soil profile need to be simplified. This is usually achieved by means of an ordination process involving the use of principal components analysis. As was discussed previously, soil properties can be considered as a series of points in multi-dimensional space. These properties can be reduced mathematically to one or a few principal axes that account for as much of the total variance as possible. The similarity of position of sites on these principal component axes allows a similarity or dissimilarity matrix to be produced. This matrix is then used in the grouping procedure.

The simplest procedure is known as the single-linkage method and is essentially a nearest neighbour method. This can be illustrated with the example of the nine sites used by Webster and Oliver (*ibid.*) in west Oxfordshire. The positions of the nine sites on two principal component axes are shown in Figure 5.6. The single-linkage method fuses the closest pair of individuals, in this case 34 and 45, then the matrix of distances is rescanned, and the second shortest distance is found. If this is between a member of the first group and a third individual, then that individual joins the group. If, as in this case, it is between two other individuals (59 and 63), these are fused to form a second group. If two closest individuals are in two groups, the two groups are fused. This process can be portrayed as a branching tree or dendrogram (Figure 5.7). Classification is produced from the dendrogram by drawing horizontal lines and noting the groups.

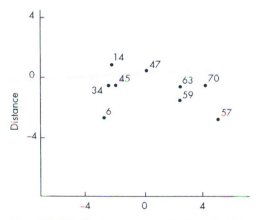

Figure 5.6 Distribution of nine soils from the Wyre Forest, England, on two principal component axes.
Source: From Webster and Oliver (1990)

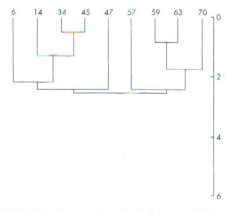

Figure 5.7 Dendrogram for a single-linkage hierarchical classification of the nine soils portrayed in Figure 5.6.
Source: From Webster and Oliver (1990)

Thus a classification with a certain number of groups can be produced.

Single-linkage is simple and easily understood, but it has disadvantages, the most significant of which is that it is prone to the process known as 'chaining'. As a group grows, it is more likely that it has neighbouring individuals that will soon fuse with it. This is what is meant by chaining. Although some properties possess a chain-like structure, with soil it is usually an artifact of the method. Chaining can lead to a grouping that fails to recognise obvious clusters. For this reason, other grouping strategies have been adopted. For a thorough analysis of these strategies, the reader is directed to Webster and Oliver (*ibid.*). As an illustration, just one of these methods, using group centroids, is examined. In this method, a newly formed group becomes a synthetic individual whose position in Euclidean space is defined by the centroid of the group. Distances between this centroid and other individuals are then used as the basis for the next stage in the grouping procedure. The process is repeated as each new group is produced. The effect of this method can be seen by considering the example shown in Figure 5.6. The groupings by the centroid method are shown in Figure 5.8. The first four groupings are the same, but the next fusion, of site 47 to the group

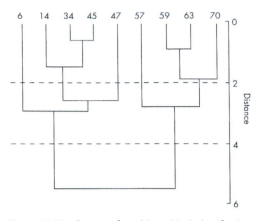

Figure 5.8 Dendrogram for a hierarchical classification produced by a centroid method.
Source: From Webster and Oliver (1990)

14 + 34 + 45, is different. Site 6 fuses at the next stage instead of before 47. This is because the centroid of group 14 + 34 + 45 has 'drifted' towards site 47 and away from site 6. This may seem a minor difference, but with larger numbers of sites it can be significant. A more realistic example of the centroid fusion of forty soils from west Oxfordshire, on each of which thirty properties have been measured, is shown in Figure 5.9.

Non-hierarchical methods

Non-hierarchical methods could be more appropriate if the properties to be classified lack an inherent hierarchical structure. As noted previously, there has been much discussion as to whether soils are linked in a hierarchical way. Many soil scientists think not, although most natural soil classification schemes are hierarchical in some respect. In non-hierarchical classification, each individual belongs to one and only one group, and any number of partitions of the population into different numbers of groups can be tried. Classes are created within which there is a minimum of variation and between which differences are maximised. Various methods of achieving this are examined by Webster and Oliver (1990).

Comparison of methods

In most cases, different results will be achieved by using hierarchical and non-hierarchical strategies on the same data. Figure 5.10 shows the optimal six groups for 201 sites in the Wyre Forest, in the English Midlands. Having established that six groups formed the optimal grouping, the dendrogram produced by a flexible sorting hierarchical strategy was cut at the six-group level. These six groups are shown in Figure 5.11; they are not well separated, and this suggests that the hierarchical method has been less effective than the non-hierarchical method in subdividing the soil profiles.

If the population consists of well-defined clusters, then both hierarchical and non-hierarchical methods should identify these clusters. If a population lacks clusters, a hierarchical method will produce a misleading picture of its structure. As Oliver and Webster (1990) note, many properties of soil are only weakly clustered and can be divided in many equally reasonable ways. Non-hierarchical classification schemes have an additional advantage in that individuals are not irrevocably assigned to groups. They can be reallocated as groups change their characteristics. Hierarchical methods do not allow this.

An interesting suggestion has been made by Webster and Burrough (1974) that combines the knowledge of the soil surveyor and numerical methods. Some soil profiles can be identified in the field as belonging intuitively and unequivocally to a particular class. These are regarded as the 'cores' of the classes. Every other profile is then compared with these cores numerically and allocated to a specific core. This appears to bridge the gap between traditional and numerical classification schemes.

SUMMARY

Soils, with their infinite variety of properties, are extremely difficult to classify, although this has not dampened the enthusiasm of numerous soil scientists and organisations in attempting to

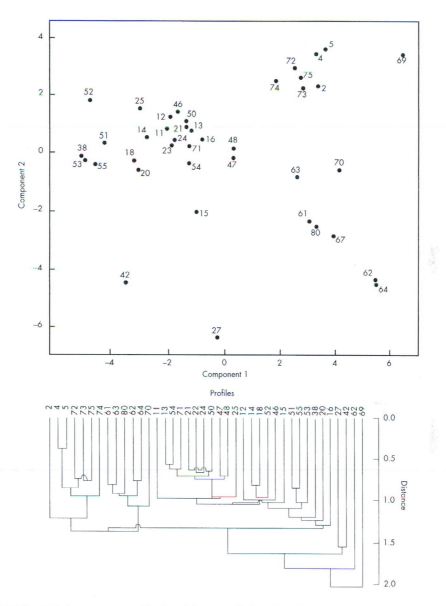

Figure 5.9 Hierarchical numerical classification of forty west Oxfordshire soils.
Source: From Webster and Oliver (1990)

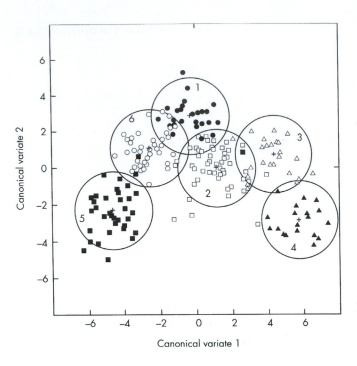

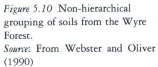

Figure 5.10 Non-hierarchical grouping of soils from the Wyre Forest.
Source: From Webster and Oliver (1990)

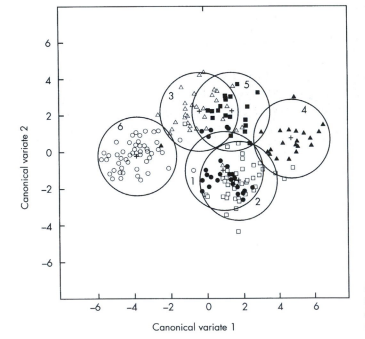

Figure 5.11 Hierarchical grouping of soils from the Wyre Forest.
Source: From Webster and Oliver (1990)

produce the most efficient division of soils. Some form of classification is need if soil maps are to have any useful purpose. The main problem concerns the production of all-purpose soil classifications. The difficulty in producing such classifications is clear from the numerous, largely different, national soil classification schemes, the most significant of which have been examined and compared in this chapter. They all rely to a greater or lesser extent on the recognition of distinctive soil properties and diagnostic horizons. They are almost all generic, hierarchical schemes. The use of hierarchical schemes has been challenged, and a number of alternative schemes have been proposed. The increased availability of high-speed computers has enabled large amounts of data to be manipulated in a manner not before possible. The results of such statistical manipulation of data are beginning to challenge the more traditional forms of soil classification.

ESSAY QUESTIONS

1 **What is the justification for thinking that some horizons (diagnostic horizons) are of more significance than others?**

2 **With reference to any ONE major classification scheme, examine the problems associated with its use.**

3 **Why may non-hierarchical soil classifications be preferable to hierarchical schemes?**

FURTHER READING

Avery, B.W. (1990) *Soils of the British Isles*, Wallingford: CAB International.

Buol, S.W., Hole, F.D. and McCracken, R.J. (1980) *Soil Genesis and Soil Classification*, Ames: Iowa State University Press.

Curtis, L.F., Courtney, F.M. and Trudgill, S.T. (1976) *Soils in the British Isles*, Harlow: Longman.

Fanning, D.S. and Fanning, M.C.B. (1989) *Soil: Morphology, Genesis and Classification*, New York: J. Wiley & Sons.

FitzPatrick, E.A. (1986) *An Introduction to Soil Science*, Harlow: Longman.

Isbell, R.F., McDonald, W.S. and Ashton, L.J. (1997) *Concepts and Rationale of the Australian Soil Classification*, Canberra: ACLEP, CSIRO Land and Water.

Soil Survey Staff (1975) *Soil Taxonomy: A Basic System of Soil Classification for Making and Interpreting Soils Surveys*, Agriculture Handbook No. 436, US Department of Agriculture, Washington, DC.

Soil Survey Staff (1992) *Keys to Soil Taxonomy*, Soil Management Support Services Technical Monograph No. 19, Blacksburg, Va: Pocahontas Press.

Webster, R. and Oliver, M.A. (1990) *Statistical Methods in Soil and Land Resource Survey*, Oxford: Oxford University Press.

6

MAIN WORLD SOILS

INTRODUCTION

The intention in this chapter is to provide a relatively comprehensive coverage of world soils. Although the FAO scheme is adopted, every attempt is made to note the equivalent soils in other classifications. In the next chapter, when considering soil patterns, it is largely the Soil Taxonomy system that is used. Thus Chapters 5 to 7 should provide a good working knowledge of the two main soil classification systems in use. The aim has also been to stress links between soils, processes and environmental factors and to assess critically the current state of knowledge. Where uncertainty still exists, this is stressed. Not all the soil types in the FAO system are covered; most attention is given to the major soil types or those that illustrate a particular principle. The Soil Taxonomy equivalents are given after the FAO soil type.

ACRISOLS (ULTISOLS)

These are acid soils (pH 5.5) with low base status. They are highly variable and are most probably polygenetic in origin, having formed in material that has already suffered weathering and some form of soil development. They all possess a marked increase in clay content with depth (an argillic B horizon), but there is usually a low frequency of clay coatings in the argillic horizon. There are five subdivisions: orthic acrisols, ferric acrisols, humic acrisols, plinthic acrisols and gleyic acrisols.

There may be a thin litter layer at the surface, but the soil surface is often bare. The upper mineral horizon is a dark greyish brown or brown sandy loam (Ah). The organic content of the Ah horizon can be as much as 10 per cent and C:N ratios of 15 indicate a moderate degree of humification. This passes quite sharply into a greyish brown or brown sandy clay loam E horizon, which changes gradually into a red clay loam argillic B horizon with some clay coatings on ped surfaces. This horizon grades into weathered rock, transported weathered soil or superficial sediments.

The soils get most of their acidity from the parent rock, being associated with acid rocks and acid sediments of Pleistocene age. The contrast in clay content between the upper and middle parts of the soils appears not to be the result of clay translocation but is more probably due to clay in the upper horizons having been destroyed and removed from the soil. Some soils are clearly old and can show an evolution back to Tertiary times. In parts of North

America, acrisols can be traced from the present land surface beneath loess as a fossil soil or soil strati-graphic unit (see Gerrard, 1992a, for a discussion of fossil soils). Climate change to cooler regimes appears to have increased acidity and led to greater hydrolysis and decomposition of clays in the Ah and E horizons.

They occur in the wet equatorial and tropical wet/dry regions of the southeastern USA, some Mediterranean countries, southeastern Australia and the humid tropics, in general where there is a heavy and continuous supply of precipitation throughout the year

ANDOSOLS (ANDISOLS)

These are soils formed on volcanic ash and other pyroclastic material that is rich in volcanic glass. They usually have dark-coloured surface horizons and possess a mollic or ambic A horizon, possibly overlying a cambic B horizon or an ochric A horizon and a cambic B horizon. There are four subdivisions: ochric andosol, mollic andosol, humic andosol, vitric andosol.

Andosols form rapidly, especially under humid conditions. Hydrolysis of the volcanic ash initially produces an amorphous yellow-brown silicate containing calcium, magnesium and potassium known as palagonite. This changes very rapidly to allophane. Humification of organic matter produces a very stable complex with allophane. Andosols possess extremely high porosities, sometimes over 70 per cent, and often exhibit fluffiness. Both of these characteristics are the result of allophane. Iron and aluminium oxides are also produced. The soils, although freely drained, also have good water retention because of the allophane.

Andosols occur in a variety of climatic environ-ments from the tundra to the tropics, but conditions must be humid. Andosols do not occur in dry climates. Volcanic landscapes tend to be very unstable, thus the best-developed soils are found on flat, stable land surfaces where volcanic activity has ceased or is limited in frequency and intensity.

Under such conditions, andosols eventually develop into other soil types. Andosols are generally characteristic of volcanic areas such as Japan, New Zealand, northwestern USA, the East and West Indies, Hawaii, East Africa, and parts of the Andes, which are producing or have produced large amounts of volcanic ash.

ARENOSOLS (PSAMMENTS)

These are weakly developed, coarse-textured soils with poor horizon differentiation. The colour varies from yellow to red, depending on drainage conditions and age. They form on sandy deposits such as sand dunes and sheets, and on river terraces. They can form on other materials and evolve from other soils where selective erosion, either water or wind, has removed the finer materials. Thus they may have developed from ferralsols. If erosion has removed the finer elements from the ferralsol, a coarse-textured soil residue will be left in which the arenosol can develop. They can also develop on the weathered residue of acidic rocks such as granite, after leaching has removed the clay from the profile. There are four subdivisions: cambic arenosols, luvic arenosols, ferralic arenosols and albic arenosols.

The upper mineral (Ah) horizon is usually relatively thin (10–15 cm) and only slightly stained by organic matter. There is usually less than 2 per cent organic matter in the upper horizon, and C:N ratios of less than 12 indicate a good level of humification. There is a characteristic single-grain structure, and the coarse texture inhibits soil organism activity. This horizon grades into a coarse sandy cambic B horizon, which may be over 2 m thick. There may also be a very slight increase in clay content at this level. They are weakly acid (pH values 6.0 to 7.0), with high base saturation, but CEC values are low because of a lack of clay. They are found mainly in tropical wet/dry, tropical desert and steppe climates in East and southeastern Africa, the southern Sahara, central and Western Australia, and parts of equatorial South America.

CAMBISOLS (INCEPTISOLS)

Cambisols are fairly uniform brown soils with a cambic B horizon and no diagnostic horizons apart from an ochric or umbic A horizon, or a calcic or gypsic horizon (Figure 6.1). They form a highly variable group with nine subdivisions: eutric cambisols, dystric cambisols, humic cambisols, gleyic cambisols, gelic cambisols, calcic cambisols, chromic cambisols, vertic cambisols and ferralic cambisols. They form best under deciduous forest (oak, beech, hazel) in maritime west coast or humid continental climates on all topography, but especially on gentle slopes and at the base of slopes where water movement brings in dissolved cations. Their main areas are in Central and Western Europe, east and central North America, and eastern New Zealand.

As an example, dystric cambisols possess a loose, leafy litter layer resting on a 5–20 cm humose, greyish brown, granular Ah horizon rich in faecal material. This grades into the brown middle Bw horizon, up to 30 cm thick, with an angular or subangular structure, the result of wetting and drying. Soils are usually of medium texture, with the maximum amount of clay in the upper horizons. Clay content might decrease with depth or stay constant. The pH in the upper horizons is 5.0 to 6.5,

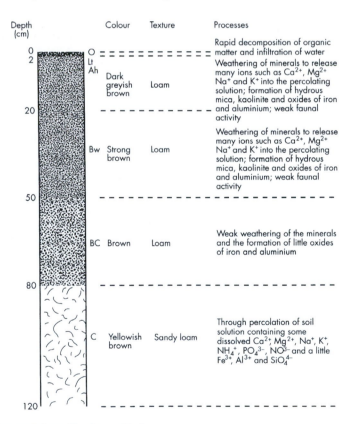

Figure 6.1 Characteristic profile of a cambisol.
Source: After FAO/UNESCO (1989); FitzPatrick (1983)

increasing to 7.0 at depth. Organic content ranges from 3 to 15 per cent in the upper horizons, with C:N ratios between 8 and 12, indicating a high state of humification. Litter is decomposed rapidly by micro-organisms and mesofauna. CEC values are 15–30 meq kg^{-1} at the surface, decreasing with depth. Calcium is the principal exchangeable cation, but magnesium may be common locally. Base saturation values are influenced by climate, being higher at the drier end of the pedogenic gradient.

Hydrolysis is an important process and is at a maximum at the surface because of the organic matter. Most of the iron and aluminium released is precipitated close by and is not removed or redistributed by leaching. Leaching of basic cations is important, however, and most of the sodium, potassium and magnesium ions released are lost completely. Calcium may be precipitated as calcium carbonate in the lower soil horizons in drier climates. Cambisols form on parent materials that are often unconsolidated and basic or calcareous. If release of basic cations does not match removal by leaching, the soil becomes more acid and may develop into a podzol or luvisol.

CHERNOZEMS (MOLLISOLS)

Chernozems are one of the most distinctive and least variable of soils. This may be due to the restricted environmental conditions within which they form and the short length of time that they have been forming. Most are less than 10,000 years old. They develop almost exclusively on loess on the flat to gently undulating continental mid-latitude steppes under tall grass vegetation. They may develop on other sediments rich in calcareous material but are poorly developed or absent on steep slopes. Their main area is south central western Russia and the former Soviet Union, although there are areas in the central and north central USA.

There are four subdivisions: haplic chernozems, calcic chernozems, luvic chernozems and glossic chernozems. Calcic chernozems are examined here (Figure 6.2). There is a thin, loose, leafy surface litter resting on a root mat, which is followed by a dark calcareous upper Ah horizon, usually 50–100 cm thick. The Ah horizon has a granular or vermicular structure and abundant earthworm casts. Calcium carbonate may be deposited in the lower part of the Ah horizon as a white thread-like pseudo-mycelium. Organic matter decreases with depth, but calcium carbonate increases and forms a calcic horizon, usually in concretionary form but sometimes massive. The lowest part of the soil is often unaltered calcareous loess. Crotovinas are common, resulting in thoroughly mixed upper horizons.

Clay is generally distributed uniformly with depth, although there may be a slight maximum between 50 and 200 cm, the origin of which is uncertain. As noted, there is a steady increase in carbonate with depth, and the surface horizons might be sufficiently decalcified to produce slightly acid conditions. Organic matter in the upper horizons ranges from 3 to 15 per cent, with C:N ratios of 8–12, indicating reasonable humification. Much of the organic matter is from decomposing roots. The pH varies from 5.5 to 8.0 in the upper horizons, depending on how much leaching and decalcification has occurred, and it may reach 8.0 or even higher just above the parent material. CEC values are low as a result of a low clay content.

The main processes involved are the rapid incorporation of organic matter, humification, and leaching of soluble salts and carbonates. Organic incorporation is achieved by earthworm activity, but churning and mixing of the upper horizons is also achieved by a variety of small invertebrates. Rainfall amounts are low, but there is considerable leaching because of the solubility of the salts. The peds can retain large amounts of water, but there is free movement of water through the well-developed structure.

FERRALSOLS (OXISOLS)

Ferralsols exhibit great variation in properties and are difficult to classify. They are the relatively old soils on stable land surfaces under wet equatorial or

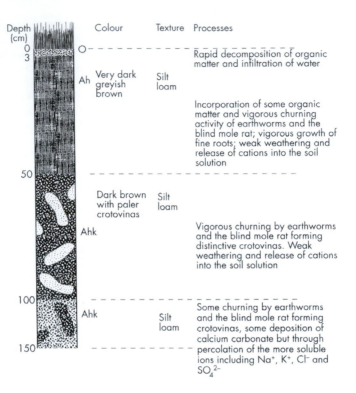

Depth (cm) | Colour | Texture | Processes

O — Rapid decomposition of organic matter and infiltration of water

Ah | Very dark greyish brown | Silt loam | Incorporation of some organic matter and vigorous churning activity of earthworms and the blind mole rat; vigorous growth of fine roots; weak weathering and release of cations into the soil solution

Ahk | Dark brown with paler crotovinas | Silt loam | Vigorous churning by earthworms and the blind mole rat forming distinctive crotovinas. Weak weathering and release of cations into the soil solution

Ahk | | Silt loam | Some churning by earthworms and the blind mole rat forming crotovinas, some deposition of calcium carbonate but through percolation of the more soluble ions including Na^+, K^+, Cl^- and SO_4^{2-}

Figure 6.2 Characteristic profile of a chernozem.
Source: After FAO/UNESCO (1989); FitzPatrick (1983)

seasonally moist tropical climates. They occur on all types of consolidated rock. The derivation of the name informs us that they are rich in iron and aluminium, and the thickness of the oxic B horizon can be over 10 m. A concretionary horizon is often present that may harden to produce a plinthite. Colour is also variable (10YR to 10R), with the most intense colour often in soils developed on material rich in ferromagnesian minerals or on limestones.

Ferralsols tend to possess a sparse litter layer, and in open forests and on grasslands there may be frequent termitaria. The uppermost mineral horizon is usually less than 20 cm thick and is a red or greyish red, pale (ochric) A horizon with slight organic staining (Figure 6.3). The surface horizon

usually has less than 5 per cent organic matter content, and C:N ratios of 8–12 indicate much humification. A predominantly granular or sub-angular blocky structure may be disturbed by earthworm and termite activity. The colour deepens with depth into the red or brownish red oxic B horizon, which is often thicker than 3 m. This horizon generally possesses a sandy clay to clay texture with massive and incomplete blocky structure. Sand grains are set in an isotropic matrix, and there may be thin clay coatings.

The oxic B horizon grades into a weathering profile that may be many metres thick. As with any normal weathering profile, the degree of hydrolysis decreases downwards. Feldspars are usually replaced by a pale yellow clay with kaolinite dominating,

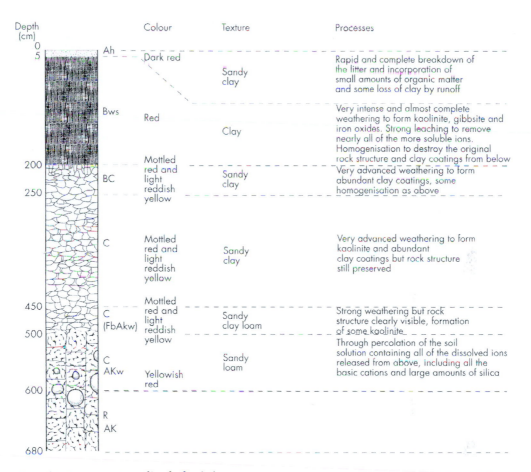

Figure 6.3 Characteristic profile of a ferralsol.
Source: After FAO/UNESCO (1989); FitzPatrick (1983)

although gibbsite and hydrous mica are relatively common. Intense weathering produces silt:clay ratios that are normally less than 0.2. With depth, the degree of alteration decreases and the amount of unaltered rock and corestones increases.

Soils are generally acid, with pH values of 5.5 at the surface, decreasing to 4.5 in the middle horizon and 5.0 in the mottled clay. CEC values are low, a result of the low activity of kaolinite. They are usually <15 meq kg^{-1} for clay and about 12 meq kg^{-1} for whole soil values. Exchangeable cations and base saturation are also low. Calcium is the principal exchangeable cation, with smaller amounts of magnesium, potassium and sodium.

Ferralsols develop by progressive hydrolysis transforming the parent material into clay, micas and oxides, with the loss in water of basic cations and silica. The specific pathway of development depends on parent material and drainage characteristics. Mottling is the result of the segregation of iron compounds, with the change in colour to reddish brown in the mottled clay suggesting less hydrated forms of goethite. The upper parts of the soils are usually freely drained, but saturated conditions may occur for short periods following heavy rain. Some of the mottled clay may be the result of this wetness.

Vegetation tends to be rainforest or semi-deciduous tropical forests on lowlands. Ferralsols are best developed on moderate slopes and are not found above about 1,200 m, where conditions are wetter and cooler. Some ferralsols are quite old, extending back to the middle of the Tertiary Period. They occur in large areas in northern Australia, much of India, Africa both sides of the equator and South America, particularly Brazil.

FLUVISOLS (FLUVENTS)

These are soils of alluvial deposits, usually on floodplains, but they can develop on marine, lacustrine and colluvial material. Because of this, they receive fresh material at quite regular intervals. They possess no diagnostic horizons, except for the possibility of a histric H horizon. Subdivision is on the basis of base saturation, whether they are calcic or not and whether there is a concentration of sulphidic material. Eutric fluvisols have a base saturation greater than 50 per cent between 20 and 50 cm from the surface and are not calcareous at the same depth. Dystric fluvisols have a base saturation less than 50 per cent between 20 and 50 cm from the surface, calcic fluvisols are calcareous between 20 and 50 cm from the surface, and thionic fluvisols possess a sulphuric horizon or sulphidic material or both less than 125 cm from the surface.

All fluvisols go through the processes known as ripening (see Chapter 3) when flooding is prevented and they dry out. Chemical ripening involves the oxidation of organic matter, iron and manganese. Biological ripening involves colonisation by vegetation: the plants extract water and aid the drying-out process. Homogenisation of the soil by roots, earthworms and other soil fauna occurs, creating passages for the ingress of air. Physical ripening includes a reduction of volume as organic matter is oxidised and water is removed. This leads to settling, cracking, shrinking and the formation of a prismatic structure. Inwash might occur, leading to soil coatings.

Thionic fluvisols are rather special soils that occur under mangroves in the estuaries and deltas of tropical rivers. A surface organic/mineral mixture overlies material with yellow mottles, predominantly of jarosite. At the base, there is a dark grey or bluish grey, completely anaerobic horizon containing pyrite. pH values are distinctive, being 4.5 at the surface but decreasing to 3.5 in the horizon containing jarosite, followed by an increase to over 5.0 in the lowest anaerobic horizon. As noted in Chapter 3, reduction of sulphates to hydrogen sulphide by sulphate-reducing bacteria occurs under anaerobic conditions. There is then reaction of hydrogen sulphide with iron to form ferrous sulphide and pyrite. If the soil remains anaerobic, the reaction stops and there is a build up of pyrite. With drainage, oxidation of ferrous sulphide occurs to form ferric sulphate and sulphuric acid. Hydrolysis then changes ferric sulphate to basic ferric sulphate (jarosite).

The sulphuric acid produces several unfortunate consequences. Acidity increases and dissociated hydrogen ions replace base cations on the exchange complex. Primary minerals are attacked, causing nutrients such as calcium and potassium to be released and lost in solution. Clay minerals are also attacked, forming aluminium sulphate and the precipitation of aluminium hydroxides. Under these circumstances, pH may drop as low as 3.0. Such is the situation in many tropical coastal areas of Southeast Asia, East and West Africa, Surinam, and tropical and subtropical parts of Australia.

GLEYSOLS (AQUEPTS, AQUENTS)

These are essentially waterlogged soils, generally on unconsolidated material, that show hydromorphic characteristics within 50 cm of the surface. There are seven subdivisions: eutric gleysols, calcaric gleysols, dystric gleysols, mollic gleysols, humic gleysols, plinthic gleysols and gelic gleysols. Only humic and gelic gleysols are examined here.

Humic gleysols are characterised by a mottled grey or olive cambic B horizon as a result of prolonged periods of anaerobic conditions in cool climates. A spongy, matted surface litter rests on well-decomposed organic matter, up to 20 cm thick, which is dense and massive and contains many fine roots. Organic matter content in the O horizon is often greater than 50 per cent, with C:N ratios over 20. There is then a sharp change to the dark grey Ah horizon (10 per cent organic matter, C:N ratios 15–20), which is a mixture of organic and mineral material, and another sharp change to the mottled grey or olive Bg horizon (organic matter less than 1 per cent). The mottling may be irregular, or it may be related to pores and ped surfaces. Clay is uniformly distributed with depth, and clay coatings are quite frequent. The lowest horizon may be uniformly grey, olive or blue because of permanent saturation. The pH at the surface is usually around 4.5 due to acid litter and decomposition under cool conditions, but it increases to about 7.0 with depth. Base saturation is usually high.

Humic gleysols are common on glacial deposits but also occur on a variety of sediments. The prime consideration is drainage; therefore topography, such as flat surfaces, depressions and lower slope portions, is more important than climate. However, they do tend to be more common in high-precipitation areas with cool climates dominated by wet-loving vegetation such as *Sphagnum*, *Eriophorum* and *Juncus*. They are generally quite young and are related to landscape age.

Gelic gleysols are formed on permafrost in areas where mean annual temperatures are below −1°C. These are the taiga areas of central Siberia, Canada and Alaska with vegetation of grasses, sedges, lichens, mosses and dwarf *Salix*. Waterlogging creates their essential characteristics, and cryoturbation by annual freezing and thawing creates many soil patterns. On slopes, horizons may be destroyed by gelifluction.

The soils possess a thin litter resting on a partially humified Oh horizon up to 15 cm thick. This is underlaid by a massive, loamy, grey cambic B horizon up to 25 cm thick with ochrous or brown mottling. As the layer of permafrost is approached, discrete areas of organic matter and mottled mineral matter may occur. The change to permafrost is usually sharp. Soils are usually loamy in texture, with a uniform size distribution with depth. The surface horizon is moderately acid, and pH increases with depth. CEC values mirror organic content, being low when organic content is low. They will be maximum at the surface, with another maximum where organic matter accumulates just above the permafrost. Base saturation is generally low but variable. Low humification rates lead to relatively high C:N ratios. Most soils have formed since the last retreat of the glaciers.

GREYZEMS (MOLLISOLS)

Greyzems appear to be intergrades between chernozems and luvisols. They develop in warm continental areas on gentle slopes beneath grassland. They possess a mollic A horizon and bleached coatings on structural peds. There are two subdivisions: orthic greyzems and gleyic greyzems.

HISTOSOLS

Histosols are peat soils with an H horizon of over 40 cm. They are subdivided into eutric histosols (pH of 5.5 or more between 20 and 50 cm of the surface), dystric histosols (pH of less than 5.5 between 20 and 50 cm of the surface) and gelic histosols, which have permafrost within 200 cm of the surface. Although they are highly organic, it is usually possible to identify specific horizons with recognisable properties. The uppermost part is composed of fibric plant remains showing little evidence of decomposition. This grades into pseudo-fibrous material, which breaks down easily when rubbed. This layer can also be quite thick and grades into black, massive, amorphous and gelatinous matter in an advanced state of decomposition. Histosols clearly need an environment where the surface remains wet and to which organic matter is added continuously. Wetness can be caused by water accumulating in a depression or due to extremely high humidity and precipitation. A basic distinction can be made between basin histosols and blanket histosols.

Basin histosols

These develop in flat, low-lying landscapes, valley floors, lagoons, and natural depressions such as in glacial or fluvioglacial deposits. They may be over 10 m thick and, as they develop from the surface, may record major climate or environmental changes that have occurred during their development. Minerals layers may indicate erosional instability on neighbouring slopes or an aeolian input. As the peat develops, it may rise above the water layer in a dome, and the vegetation will gradually change to drier species. pH values will depend on the composition of the water. Acid crystalline rocks will produce very acid conditions, whereas where basin histosols develop on calcareous parent materials, such as in the fens of eastern England, higher pH values are found.

Blanket histosols

These develop in areas with high humidity and high precipitation levels where rates of organic decomposition are low. They cover more than 150 million hectares of the Earth's surface, with over 5 million hectares in areas such as the muskeg of Canada, the British Isles, Finland, Germany, the USA, and Russia and the former Soviet Union. Blanket histosols are always acid, because the water comes directly from precipitation. Organic matter tends to be more decomposed, because water and dissolved oxygen are moving continuously through the material. Blanket histosols rest on bedrock or old soil and are rarely more than 3–4 m thick. In northern Europe, the major phase of blanket histosol development started about 7,500 years BP. If climatic conditions are favourable, they may occur on steep slopes, but they are easily eroded and may move downslope as peat flows or bog bursts.

KASTANOZEMS (MOLLISOLS, CHESTNUT SOILS)

Kastanozems are soils that grade to chernozems in one direction and to solonetz and solonchaks in the opposite direction. At the surface they possess a loose, leafy litter resting on a dark brown, granular A horizon about 50 cm thick. The upper part possesses abundant fine roots. With depth, the structure changes from fine prismatic to more massive, and there may be some accumulation and concretions of carbonates. There is usually an accumulation of gypsum between $c.$ 1 and 2 m. Texture is uniform with depth, usually in the silt range as a result of predominantly loessic parent material. There are three subdivisions: haplic kastanozems, calcic kastanozems and luvic kastanozems.

Organic content ranges from 3 to 6 per cent in the upper horizon, and C:N ratios of 8–12 indicate a high degree of humification. The pH is usually 7.0 at the surface but increases to over 8.0 in the zone of carbonate accumulation. CEC values in the upper horizon are 20–30 meq kg^{-1}. The soil is usually saturated with basic cations, with calcium predominant, followed by magnesium, potassium and sodium. Soil organism activity is high, as are leaching quantities, and ions of sodium, chloride and sulphate tend to be removed completely from the soil. Calcium is translocated as bicarbonate and deposited as calcium carbonate in the lower part of the soil.

Some kastanozems may be polygenetic and may originally have been luvisols. Certainly some kastanozems (the reddish chestnut soils of the south central USA) seem polygenetic and can be traced to the Sangamon Interglacial fossil soil. They occur in the mid-latitude steppes of the mid-western USA and southern Russia, where climate is seasonally dry or semi-arid. The vegetation is medium-height grassland, and the topography is flat to gently undulating. Most are less than 10,000 years old.

LUVISOLS (ALFISOLS, BROWN EARTHS)

Luvisols are most commonly associated with deciduous woodland, especially of oak and beech, but are sometimes found under mixed deciduous/coniferous forest. They occur very occasionally under grassland. Most appear to have developed in the Holocene, although some are undoubtedly older. They develop best on flat to gently sloping topography. Their main characteristic is an argillic B horizon with a base saturation of at least 50 per cent. There are eight subdivisions: orthic luvisols, chromic luvisols, calcic luvisols, vertic luvisols, ferric luvisols, albic luvisols, plinthic luvisols and gleyic luvisols.

At the surface there is a loose, leafy litter resting directly on the mineral soil or on a thin layer of humified plant remains (Figure 6.4). This thin layer passes rapidly into a dark grey or greyish brown granular ochric A horizon about 10 cm thick, which is a mixture of mineral grains and fragments of humified organic matter. Organic matter content in the upper horizons is 5–10 per cent, and C:N ratios of 12–18 reflect moderate humification. Throughout the soil, organic matter and C:N ratios are low, indicating an advanced state of organic decomposition.

Luvisols appear to be formed by the progressive movement of material down through the soil. Initially, soluble salts and carbonates would be moved, but this would be followed by clay. Some of this clay is redeposited in the middle horizons; however, it is possible that some clay is removed from the soil completely. Free oxides of iron, aluminium and silica are moved at the same time as the clay. The processes seem to operate best in climates with a distinct dry season, such as tropical wet/dry, mediterranean and humid continental areas. There is translocation of material during the wet season, followed by partial dehydration during the dry season and the attachment of particles to peds. Continual repetition of this process will lead to the formation of coatings.

NITOSOLS (ALFISOLS, ULTISOLS)

These are strongly weathered kaolinitic soils with shiny ped surfaces. There is a steady increase in clay content until the argillic B horizon. Clay content then remains relatively uniform until less weathered material is reached. Clay translocation does not seem to be occurring – the clay content differences are the result of progressive weathering and hydrolysis in the upper horizons reducing the clay content. Nitosols are common in tropical and subtropical areas, and there are three subdivisions: eutric nitosols, dystric nitosols and humic nitosols.

PHAEOZEMS (MOLLISOLS)

Phaeozems possess a mollic A horizon and lack most of the diagnostic horizons that make other soils distinctive. There are four subdivisions: haplic phaeozems, calcaric phaeozems, luvic phaeozems and gleyic phaeozems. The main type of phaeozem, luvic phaeozems, is examined here; they tend to form a continuum between chernozems and luvisols. They occur mostly on unconsolidated parent material of glacial drift, loess and sometimes alluvium on flat or gently sloping surfaces. They are generally associated with tall grass prairie in North America, but here soil and vegetation may be out of phase. In Europe, they are often found beneath deciduous woodland.

At the surface, there is usually a thin, loose leafy litter resting on a mineral soil or on a thin root mat. The upper mineral horizon is a very dark grey A horizon with a granular or vermicular structure (Figure 6.5). As with chernozems, this horizon can be quite thick, often up to 50 cm. This upper horizon grades into a dark brown, angular/subangular blocky argillic B horizon with clay coatings on ped surfaces. The soil, with depth, grades into unaltered

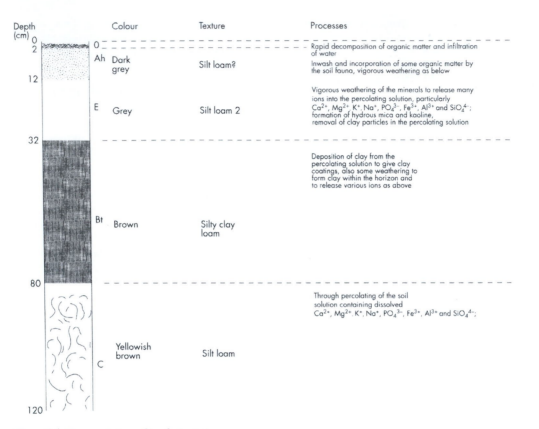

Depth (cm)	Colour	Texture	Processes

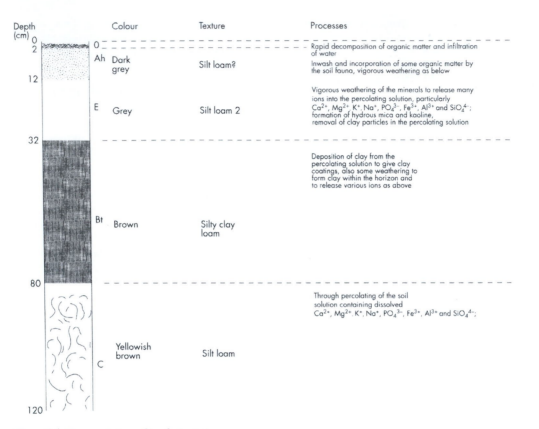

Figure 6.4 Characteristic profile of a luvisol.
Source: After FAO/UNESCO (1989); FitzPatrick (1983)

material, and the clay content and clay coatings decrease.

The soil is generally acid, although the pH at the surface may be >7. Organic matter is approximately 5 per cent in the top layers but decreases to 1–2 per cent in the middle layers. This is a comparatively high figure for a middle horizon and explains the dark colour. The organic matter in the upper horizons is well humified, which explains C:N ratios between 10 and 12. CEC values are extremely variable, depending on clay content and the specific clay mineralogy. If montmorillonite is present in appreciable amounts, the value is about 25–30 meq kg^{-1}, whereas where mica is predominant

values are lower. Calcium is the dominant exchange ion, and base saturation is normally above 80 per cent. Lowest values of base saturation tend to occur in the middle horizons, where the pH is lowest.

As luvic phaeozems appear to form a continuum between chernozems and luvisols, this suggests that they may not be in phase with current environmental conditions. Many appear to have developed in two phases. In central North America, there is abundant evidence for successive phases of vegetation colonisation in step with Pleistocene climatic changes, especially those following the last glacial maximum. Spruce forests have been followed by a predominantly hardwood vegetation and then

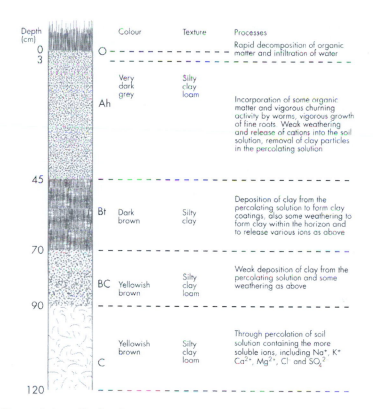

Figure 6.5 Characteristic profile of a phaeozem.
Source: After FAO/UNESCO (1989); FitzPatrick (1983)

prairie grassland. The first stage in soil formation seems to have been the development of luvisols, which could account for the middle horizon of clay maximum. The clay content in the argillic horizon can be 30–40 per cent. The second stage would have involved the superimposition of an organic upper horizon as a result of the prairie vegetation. Thorough mixing by soil organisms would have occurred. However, in Russia and the former Soviet Union, luvic phaeozems occur under deciduous woodland and still possess thick organic horizons. This implies that the organic layer is not necessarily due to prairie grass vegetation. What appears to be certain is that the middle, clay-rich, horizon is a relic from the operation of previous soil processes.

Phaeozems occur in essentially continental climatic conditions, but with wide variations in temperatures (2–18°C mean annual temperatures) and precipitation amounts (400–1,200 mm). However, the two climatic parameters are in phase and, because evaporation decreases as temperature decreases, similar amounts of moisture are passing through the soils.

PLANOSOLS (ALFISOLS, ULTISOLS, ARIDISOLS)

These are soils developed on level topography with poor drainage, possessing an albic E horizon over a slowly permeable horizon. They show hydro-morphic properties in at least part of the E horizon. They appear to develop from gleysols. There are six subdivisions: eutric planosols, dystric planosols, mollic planosols, humic planosols, solodic planosols and gelic planosols.

Eutric planosols (albaqualfs) show a marked increase in clay content in the mottled, brown, grey or olive argillic B horizon. The change in texture is the result of *in situ* weathering under wet conditions in the middle horizon and the destruction and removal of clay from the albic E horizon. Wetness is created by depressions, an impermeable horizon or a high clay content in the parent material, which inhibits water movement.

Solodic planosols (natraqualfs) are strongly leached solonetz, which causes them to become progressively more acid. The albic E horizon has a pH of 5.0 and less than 30 per cent base saturation, and the natric B horizon (original pH 8.5 or more) may have a lower pH and will have lost some of the high content of exchangeable sodium and magnesium.

PODZOLS (SPODOSOLS)

There are six subdivisions of the podzol group: orthic podzols, leptic podzols, ferric podzols, humic podzols, placic podzols and gleyic podzols. The two most important, orthic podzols and placic podzols, are discussed here.

Orthic podzols (orthods)

Orthic podzols are found in a large part of the taiga and tundra zones of northern Asia, Europe, North America and New Zealand under coniferous forest or heaths, vegetation that produces acid litter, and on topography that varies from gentle to steep. Climates are mostly humid continental, with cool summers and annual precipitation amounts ranging from 450 to 1,250 mm. Orthic podzols also occur in some humid tropical areas, such as Borneo, Brazil, Malaysia and Ghana, where the material is highly siliceous.

The main characteristic of podzols is an upper pale grey, strongly leached horizon overlying a brown to very dark brown horizon with accumulations of iron, aluminium and/or humus (Figure 6.6). They possess a loose and spongy surface litter, 1–5 cm thick, followed by a partly humified organic (O1, fermentation) layer. This is of similar thickness to the litter layer, with plant fragments still visible even though decomposition is well advanced. The level of decomposition increases with depth, and a black, amorphous (O2, humifon) layer, with little evidence of plant fragments, is produced. There is then a sharp change to a very dark grey mixture of organic and mineral matter. The mineral matter is mainly quartz bleached by the acid conditions, but many of the grains possess black or dark brown coatings. This horizon grades rapidly into the bleached pale grey E horizon, with a single-grain or alveolar structure. This is followed by a very dark brown accumulation of humus, iron and aluminium in a Bs or Bhs horizon. Structure ranges from subangular blocky to very hard and massive. Clay is arranged as granules or as thin coatings around sand grains. As one proceeds further down the profile, organic content decreases and the colour changes to a strong brown. This is also a type of Bs horizon. There is then a gradation to relatively unaltered parent material, although there may be a hard, compact fragipan at depth.

The clay content in orthic podzols is low, usually <10 per cent in the bleached horizon with a slight increase in the middle horizon due to leaching and deposition of aluminium and iron hydroxides rather than the translocation of discrete clay particles. Organic matter distribution within the profile exhibits two maxima: >70 per cent at the surface and a lesser maximum in the middle horizon, where accumulation is again the result of leaching. C:N ratios follow a similar trend, being 25–30 in the surface horizons (lack of decomposition), 10–15 in the bleached horizon and 15–25 in the middle horizon. CEC values also follow the maxima of organic matter, because of little influence from clay. Exchangeable cation content is low, with the greatest amount in the decomposing organic matter,

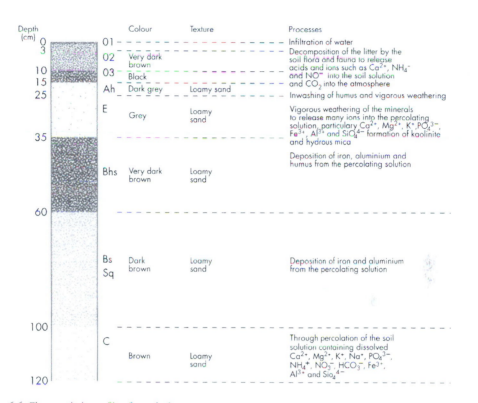

Figure 6.6 Characteristic profile of a podzol.
Source: After FAO/UNESCO (1989); FitzPatrick (1983)

where humification releases basic cations. Base saturation is low in all horizons. Acid parent material, acid litter and removal of bases by leaching produce very acid conditions in the upper soil, with pH values of 3.5 to 4.5. This increases to a maximum of about 5.5 in the relatively unaltered parent material.

Maximum free iron oxide occurs in the middle Bs horizon, with a minimum level in the upper E horizon. There is also an increase in aluminium in the Bs horizon, and Al_2O_3/Fe_2O_3 ratios show that there is more aluminium than iron, but ferric hydroxide possesses a strong colour and is more obvious.

The formation of orthic podzols is the result of a number of interacting factors and processes.

Decomposing organic matter plus water charged with carbon dioxide cause the soil solution to be acid. This causes weathering of primary silicates, the release of cations and the formation of a mobile complex of iron and aluminium. This leads to bleaching in the upper part of the soil. Most basic cations are washed through the soil system. Some silica is also lost, but some may be deposited in the bleached horizon. Some aluminium and iron is also lost, but most is deposited together with some humus in the middle layers. Processes involved in this accumulation have been examined in Chapter 3.

Clays are also formed: kaolinite in the upper horizon and vermiculite in the middle horizon. Soils are freely drained in their upper and middle horizons, aided by parent materials that are nearly always medium- to coarse-textured, with a high quartz and stone content.

Placic podzols (placorthids, iron pan podzols)

The distinguishing feature of placic podzols is a thin, hard continuous iron pan, which is comparatively impermeable (Figure 6.7). They form on coarse-grained usually siliceous material in cool, wet maritime conditions under heath vegetation in northwest Europe, Scandinavia, northern Germany, northern France, the British Isles, eastern Canada and Alaska. But they also occur in some oceanic tropical areas at high altitudes in Malaysia and the Solomon Islands. In this case, the vegetation is usually evergreen upper montane forest. They form on quite steep as well as gentle slopes.

Placic podzols possess a thin surface litter overlying a black, plastic, organic horizon, which can be up to 30 cm thick. There is then a sharp change to a greyish brown, loamy Eg horizon about 10–15 cm thick, with ochrous mottles. The thin iron pan (placic horizon) then occurs, dark brown at the top, becoming yellowish brown below. It is usually wavy and irregular, sometimes avoiding stones but sometimes going straight through them. There may be a thin root mat on top of the iron pan. Occasionally, there is more than one pan in a soil. In thin section, the pan is seen to be composed of dense and isotropic thin alternating bands. Horizons below the pan are variable and appear to relate to an earlier soil

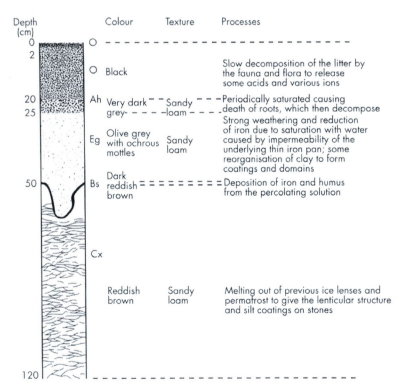

Figure 6.7 Characteristic profile of a placic podzol.
Source: After FAO/UNESCO (1989); FitzPatrick (1983)

phase. There is often a spodic B horizon and a fragipan.

It is clear that the iron pan has formed in a fully developed soil, but how is still unclear. One explanation is that as the soil solution percolates through the soil more and more iron is chelated by the organic matter in the solution until a critical concentration is reached and precipitation of iron and organic matter takes place (FitzPatrick, 1983). If this occurs at the same point in the soil each year, a pan might form. But it cannot explain the pan cutting through stones. Once it has formed, the pan creates anaerobic conditions and the reduction of iron compounds to produce olive and grey colours. Little alteration now takes place in lower horizons. This inhibition of moisture movement might allow a histosol to develop.

The soils, in general, are of medium to coarse texture, with abundant stones. They are very acid, with pH values of 3.5 at the surface and 5.0 at depth. C:N ratios are large, indicating the low degree of humification. Exchangeable cations are dominated by calcium, and the percentage base saturation is low. The iron pan contains most iron, with 10 per cent more iron and 15 per cent less aluminium than in the horizon below. Aluminium tends to accumulate in the fragipan. The fragipan, formed of lenticular compacted material, reflects former periglacial conditions and possibly permafrost. The settling of stones on thawing creates spaces for colloidal silt and clay material and aluminium hydroxide to be washed in, thus preserving the lenticular structure.

FitzPatrick (1983) has suggested that placic podzols seem to be part of a developmental sequence under cool oceanic conditions from gelic gleysols to orthic podzols (in the early Holocene) to placic podzols in the middle to late Holocene.

PODZOLUVISOLS (ALFISOLS)

These soils have developed on medium- to fine-textured material under cool, moist continental conditions in Central and Eastern Europe, central USA, and parts of Australia and New Zealand on flat or gently sloping topography where moisture can accumulate in the upper parts of profiles. The parent material restricts drainage, but there is sufficient leaching to remove carbonates and reduce base saturation. The albic E horizon often interdigitates with the argillic B horizon. These zones where the E horizon penetrates the B horizon may represent drying cracks or may reflect former ice wedges when the area experienced a periglacial climate. These are zones of preferential water movement. The soils are moderately acid and contain mica, kaolinite and vermiculite.

There are three subdivisions: eutric podzoluvisols, dystric podzoluvisols and gleyic podzoluvisols. Gleyic podzoluvisols are formed by accumulation of water during winter through to early spring, causing reduction of iron. Much of the iron and manganese is precipitated to form small concretions. Clay coatings on the faces of large structural units suggest that there is some redistribution of clay.

RANKERS (LITHIC INCEPTISOLS)

These are shallow soils on steep slopes in siliceous rocks. They tend to be mainly a variety of upper horizons resting on unaltered material. Many occur on recently deglaciated rock. Variability is caused by a variety of rock types, topography, climate and vegetation. If they develop on a thin glacial drift layer, formation may be rapid at first in the drift layer but slows as soon as the unaltered rock becomes part of the system. They occur in all parts of the world but especially in mountain areas, where soil is thin and rock slopes plentiful.

REGOSOLS (ORTHENTS, PSAMMENTS)

These soils develop on unconsolidated material and are defined by what they are not. They lack the characteristics of vertisols, aridisols or arenosols. They often form the initial stage in the development of a number of soils, such as podzols, luvisols, cambisols, chernozems, kastanozems and xerosols. Their only basic characteristic is an ochric A horizon, but this soon changes as the soil develops. There are four subdivisions: eutric regosols, calcic regosols, dystric regosols and gelic regosols.

RENDZINAS (RENDOLLS)

Rendzinas are the classic soils developed on limestone, chalk and other calcareous material. Climate is not a controlling factor; it is the parent material that is the dominant influence, although some of the characteristics of the surface horizons may vary with climate and vegetation. They occur throughout the world, and on stable land surfaces, such as in the tropics, they can be extremely old. However, many rendzinas have been developed over the last 10,000 years. Vegetation is composed of lime-loving species, and there is often a rich flora. Topography is not a major influence; rendzinas occur on steep slopes as well as on gentle topography. Soils tend to be very shallow, with a medium to fine texture and a good structure. This results in rapid water movement and, although precipitation usually exceeds evaporation, leads to them experiencing moisture stress.

There tends to be a loose, leafy litter layer resting on a dark brown or black calcareous–organic–mineral matter mixture with a well-developed crumb, granular or vermicular structure and abundant earthworm casts. There is a sharp change to underlying rock. The A horizon often contains up to 80 per cent calcium carbonate, with a pH of over 8. Organic matter content ranges from 5 to 15 per cent, and C:N ratios of 8–12 attest to the advanced state of humification. CEC values can be as high as 50 meq kg^{-1}. There is complete base saturation, with the main cation being calcium, or magnesium where the soils are developed on dolomite.

The main process is the solution and removal of carbonate, leaving a small residue of insoluble material. This residual material is mixed with humifying organic matter to produce a granular structure. The dark colour is caused by the calcareous–humus complex similar to that found in chernozems. If the insoluble residue builds up to a threshold depth, decalcification occurs, the pH falls and the soil changes in character.

SOLONCHAKS (SALORTHIDS)

These are highly saline soils that occur mostly in the drier, central parts of continents on unconsolidated material, usually loess, alluvium or material washed across pediment surfaces. They are best developed in depressions, on the beds of old lakes and in mountain basins, where water, shed from surrounding slopes, can accumulate to create waterlogged conditions. This water also brings in dissolved salts with it. They are essentially soils of the mid-latitude and tropical arid and semi-arid areas, where evaporation exceeds precipitation. However, precipitation, when it occurs, must not be sufficient to wash out the salt content of the soil. Vegetation cover is highly variable and tends to depend on the salt content of the soils. Where the salt content is over 0.5 per cent, only halophytic vegetation can survive. They are mostly Holocene in age, but many have developed more recently as a result of inappropriate irrigation (see Chapter 8).

The greatest salt content in solonchaks is at the surface, and there may be a thin surface salt crust. Salt decreases with depth. The most common ions are chloride, sulphate, carbonate, bicarbonate, sodium, calcium and magnesium, but ionic variability is extremely high. The structure is usually massive throughout. The whole soil is usually grey or greyish brown, often with mottling, which is usually greatest in the middle part of the soil (Figure 6.8). There is only a weak contrast between horizons. The upper horizon may be slightly darker as a result of staining by organic matter and can be massive, coarse platy, puffy or crusty. Varying amounts of small carbonate concretions usually occur, with ped surfaces outlined by fine carbonate crystals.

Solonchak formation is intimately related to the fluctuating position of the water table. During the

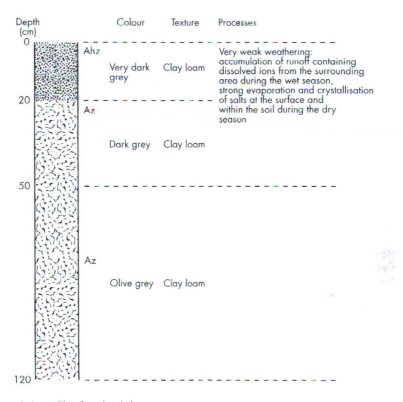

Depth (cm)		Colour	Texture	Processes

Figure 6.8 Characteristic profile of a solonchak.
Source: After FAO/UNESCO (1989); FitzPatrick (1983)

dry season, the water table is usually less than 3 m from the surface, while during the wet season it often reaches the surface, causing the reduction of iron compounds formed by the oxidation of reduced iron. When the water table drops, evaporation causes the salts dissolved in the groundwater to be deposited on the surface and in the upper soil. Continuous repetition of this cycle causes salts to accumulate within the fluctuating moisture zone, especially near the surface. There will be no salt accumulation in the zone of permanent saturation. If the water table does not reach the surface, the maximum salt content will be at some intermediate depth. Salt content will vary throughout the year as a result of some downward leaching of salt following precipitation events. Salt content will be highest following dry periods. Texture affects water retention, therefore salt contents are higher in fine-textured soils.

There is some mystery as to the derivation of the high concentrations of salts in the groundwater. Much is undoubtedly derived from the weathering of bedrock, but this process cannot account for the high amounts of chloride and carbonate. Chloride is not a normal component of rocks and is usually thought to come from seaspray or salt particles transported by the wind. Sea water incursion into groundwater systems in coastal areas is also a possibility. It is also difficult to account for the high concentrations of carbonates, especially that of sodium carbonate. Sodium is derived from rock by the hydrolysis of orthoclase feldspars or by solution from sedimentary rocks. Carbon dioxide from the atmosphere forms carbonic acid in the soil, which reacts with sodium to produce sodium bicarbonate, which is readily transformed into sodium carbonate. There are four subdivisions: orthic solonchaks, mollic solonchaks, takyric solonchaks and gleyic solonchaks. Soils with takyric features crack into polygonal elements when dry.

SOLONETZ

These are highly saline soils rich in sodium and possessing a natric B horizon but lacking an albic E horizon. They occur in semi-arid regions where there is sufficient precipitation to cause leaching of the upper horizons but where the amount of moisture is insufficient to reduce the salinity in the lower horizons. They are formed mostly in unconsolidated material, such as loess, colluvium and glacial deposits, in Australia, western Pakistan, North and southern Africa, the coastal belt of northern Chile, and parts of Argentina. In North America, they occur in patches from Texas to northern Alberta. They will not form in material containing calcium carbonate, because calcium displaces sodium and magnesium. However, they may form in such material following decalcification. They form best where the topography is flat or gently sloping but not in major depressions where the water table is close to the surface. It is more likely that solonchaks would be found in such locations. Some solonetz soils are comparatively old, but the majority have formed during the last 10,000 years.

At the surface, the soils may possess a thin and loose leafy litter resting on black humified material about 2–3 cm thick. This is followed by a brown granular ochric A horizon up to 15 cm thick (Figure 6.9). There is then a sharp change to a mottled greyish brown and brown natric B horizon, with a higher clay content and with a prismatic or columnar structure. The B horizon grades with depth into a more mottled and massive saline horizon. There may also be an intervening gypsic horizon.

The organic matter content in the surface horizons is usually less than 10 per cent, and the C:N ratio, usually less than 12, indicates a high degree of humification. pH varies from 6.0 to 7.5 at the surface but rises to 8.5 in the lowest, saline, horizons. CEC varies with texture and clay mineral type but is usually in the range 15–35 meq kg^{-1}. Apart from the upper horizons, the entire soil is saturated with basic cations. In the upper horizons, calcium is the principal exchangeable cation, but in the natric B horizon sodium and magnesium predominate. At depth, calcium may again be the principal exchange cation.

The upper horizons are usually non-saline, but salinity increases with depth. The electrical conductivity of the natric B horizon often attains 2.0 mmho cm^{-1}, but the underlying horizons may have values as high as 15 mmho cm^{-1} where there is >0.5 per cent salts and sometimes much carbonate. The amount and type of salts vary from place to place, but usually calcium, magnesium, sodium, carbonate, bicarbonate, chloride and sulphate predominate. The clay type is inherited from the parent material and is generally dominated by micas, but kaolin and smectite may be important locally.

One of the main characteristics of the soil is an abrupt and large increase in clay content on passing from the A horizon to the natric B horizon. This can be a threefold increase with most of the increase being in the fine clay (<0.2 μm) fraction. This increased clay content is difficult to explain. It was generally thought that solonetz evolved from the progressive leaching of solonchaks that were deficient in calcium ions but possessed large quantities of sodium ions. The sodium would displace the clay which migrated from the upper to the middle horizons to form the natric B horizon. But there are a number of objections to this idea. First, the increase in clay content over that in the parent material cannot be attributed solely to the small amount of clay coatings. Second, the amount of clay coatings cannot account for the amount of clay that appears to have been lost from the upper horizons.

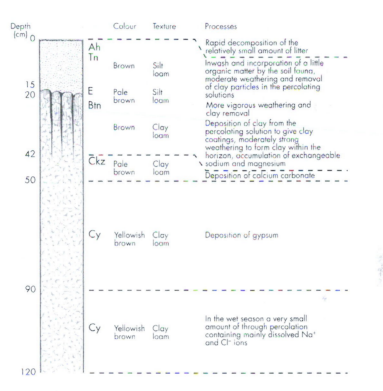

Figure 6.9 Characteristic profile of a solonetz.
Source: After FAO/UNESCO (1989); FitzPatrick (1983)

The contrast seems to be the result of processes that destroy clay in the upper horizons and create it in the middle horizons. There is also a small amount of clay translocation, which accounts for the clay coatings. The destruction of clay in the upper horizons is probably due to the dispersive action of sodium-rich waters and its increase in the B horizon to *in situ* weathering. There are three subdivisions: orthic solonetz, mollic solonetz and gleyic solonetz.

VERTISOLS (VERTISOLS)

These are very distinct fine-textured, dark-coloured soils that have quite often simply been called black earths. The name originates from the Latin *ventro*, meaning turn, and indicates that they are soils in which turning and churning is characteristic of at least the surface horizons. They form in highly weathered material derived from basalt, shale, limestone and volcanic ash and are aided by high contents of plagioclase feldspars, ferromagnesian minerals and carbonates. These minerals produce cations that enable swelling clays to form. They also form in fine-textured superficial deposits such as alluvium, lacustrine material or reworked colluvium.

They have a high clay content of mostly swelling clay minerals, especially montmorillonite. These clays swell on wetting and contract and crack when dry, creating the turnover in the top soil. Cracks are at least 1 cm wide and often extend up to 50 cm into the soil. The continuous swelling and contraction produces a number of distinctive structural features such as intersecting slickensides, wedge-shaped or parallelepiped structural aggregates and a surface gilgai microrelief. The characteristic morphology is simply the result of the swelling and contraction of the mixed layered clay minerals. Changes in volume can be as much as 50 per cent. This continual movement also produces a density of $1.8–2.0 \, g \, cm^{-3}$, which is higher than most soils. As the soil dries, some surface material falls into the cracks, which increases the pressure when the soil expands on wetting. This leads to an upward movement, which causes the churning of the soil, and the movement also produces the wedge structures and slickensides. For this to have maximum effect, there must be a dry period and a period, however short, of complete saturation. They form best in arid and semi-arid climates because of the need for alternate wetting and drying and also because some swelling clays form best under such conditions.

The surface is usually bare, although there may be a sparse litter. The uppermost middle horizon is usually thin, with a granular structure and what has been called a 'self-mulching surface' because of the expansion and contraction. In the dark middle horizon, the structural units become progressively larger with depth and with slickensided surfaces. Change to the underlying material may be gradual, by partial mixing or by interdigitating. Therefore, it may be difficult to locate the boundary, especially if the underlying material is highly weathered. It is not clear what causes the dark colour. The amount of organic matter is not high, but there may be a dark-coloured complex of organic matter and montmorillonite. Iron and manganese compounds may also be present.

Organic content is usually 1–2 per cent at the surface but can be as high as 5 per cent, and C:N ratios are usually 10–14. The high clay content ensures high CEC values of $25–80 \, meq \, kg^{-1}$. The base saturation is not usually less than 50 per cent. Most, but not all, vertisols contain free calcium carbonate as powdery deposits or concretions. Exchangeable sodium, at 5–10 per cent, is higher than for most soils but considerably less than for saline or alkaline soils. Salinity is low because of a self-flushing system. Salts tend to accumulate on the surface of the structural units in dry periods, but they get washed away by rapid water movement down the cracks during wet periods. pH values range from 6.0 to 8.5 and may increase as more sodium appears.

A number of climatic regimes produce suitable conditions, including maritime west coast, humid continental, tropical wet/dry and tropical grassland. They are found most extensively in tropical and mid-latitude desert and steppes with precipitation of 250–750 mm per year and a dry season of four to eight months. In the slightly more humid areas, the content of soluble salts and exchangeable sodium is lower, and in the more arid areas there is a gradual increase in exchangeable cations, especially calcium carbonate, and there may be a calcic horizon. Gypsum may also occur in lower horizons. Gilgai surface relief does not occur in all areas but is especially common in Australia and Texas and less common in India and Africa. Vertisols never form on slopes steeper than 8°. There are two subdivisions: pellic vertisols and chromic vertisols based on soil chroma.

XEROSOLS (ARIDISOLS)

These are soils of very dry areas that lack high salinity. They possess weak ochric A horizons plus one or more of a cambic B horizon, argillic B horizon, calcic horizon and gypsic horizon. Xerosols grade to luvisols, kastanozems, solonetz and solonchaks.

Upper horizons are usually quite thin, are often loamy and calcareous and form a yellowish red to greyish brown weak ochric A horizon. This has a platy or massive structure and grades rapidly into a calcic horizon up to 1 m thick followed by a gypsic or salic horizon that may be over 3 m thick. Organic content is low (1–2 per cent), and C:N ratios are often less than 8.

There tends to be a slight increase in clay content in the B horizon, and the carbonate content may be 12–15 per cent in the calcic horizon. This slight increase may be deceptive. The surface in extremely arid areas is subject to deflation and also rapid slopewash, and the fine surface material will be removed by these processes, leading to a concentration of coarse material at the surface. Some of the textural difference might be due to *in situ* weathering. pH values are 7.0 to 8.0 at the surface, reaching a maximum of 8.5 in the calcic horizon before decreasing with depth. CEC values are of the order of 10–15 meq kg^{-1} in the A horizon, decreasing to 8–10 meq kg^{-1} at depth. There is complete saturation of the exchange complex, and calcium is the dominant ion.

Xerosols form very slowly because of the limitation of water availability. Also, only a small amount of organic matter is produced by the sparse vegetation, and this is quickly humified. The easily soluble salts are moved from the upper part of the soil and deposited lower down. Many xerosols are relatively young, although some may extend back into the Tertiary Period. They exhibit considerable variability as a result of variations in parent material, climate and possibly erosion. The major contrast is the presence or absence of carbonates. Where there are no carbonates, the soils may be mildly acid. Xerosols are common in southern Russia and the former Soviet Union, south and north of the Sahara Desert, parts of Iran and western Pakistan, Afghanistan, and Australia. They are commonest on flat or gently sloping topography because on steeper slopes erosion removes material and lithosols, regosols and rankers are more common.

SUMMARY

All that has been attempted here is to provide a general account of the main soil types. Soil classifications are subject to continual modification. The FAO system has recently been updated with minor modifications as Technical Paper 20, Wageningen: International Soil Reference and Information Centre.

ESSAY QUESTIONS

1 How does the FAO/UNESCO classification scheme differ from the Soil Taxonomy scheme?

2 What are the main problems with using the FAO/UNESCO scheme in a field situation?

FURTHER READING

FAO/UNESCO (1974) *Soil Map of the World: Volume 1, legend*, Paris: UNESCO.

FAO/UNESCO (1989) *Soil Map of the World: revised legend*, Wageningen: International Soil Reference and Information Centre.

FitzPatrick, E.A. (1983) *Soils, Their Formation, Classification and Distribution*, 2nd edn, London: Longman.

7

SOIL PATTERNS

INTRODUCTION

Variability is an intrinsic characteristic of soils, and patterns may be identified at a variety of spatial scales. Explanations for soil patterns will differ depending on the scale of interest. At the global and continental scale, soil patterns can often be related to global patterns of climate. At a regional level, geological differences may be the determining factors. Over smaller areas other factors, such as topography and landform type, will be important. There have been many attempts to subdivide the world into homogeneous units at a variety of spatial scales. Fenneman (1916) recognised continents, divisions, provinces and sections. Divisions are subcontinental in dimension and can be broken down into provinces such as the central lowlands of North America, followed by sections, such as the Black Hills of Dakota. Southwest England would be part of the oceanic uplands province of Europe, which would include Ireland and Brittany, France. Southwest England, on its own, would be a section. For smaller units, Linton (1951) added tract, stow and site. The uplands of Dartmoor would be a tract, and valley side slopes in a particular drainage basin would be a stow. An individual valley side or an individual landform would be a site. Soil patterns at these varying scales and determining factors are now examined.

GLOBAL AND CONTINENTAL SCALE

As mentioned, climate is probably the predominant factor in causing variations in soils at global and continental scales. The general detail on the soil map of the world devised by Soil Survey Staff (1975) would suggest this (Figure 7.1). There is a general tendency for broad east–west patterns to occur approximately in accord with the main climatic groupings. This is best seen in Africa and Eurasia. The pattern in Australia tends to circle the dry interior and is again climate-related. The east–west pattern in North and South America is distorted by the major mountain ranges. Soil patterns tend to be more north to south but are still related, in a general way, to broad climatic zones.

These general relationships have been used to produce broad soil patterns on an imaginary supercontinent (Figure 7.2). Using Soil Taxonomy terminology, high northern latitudes are dominated by inceptisols, entisols and spodosols. The characteristics in common are low additions of organic matter and low organic and mineral transformations,

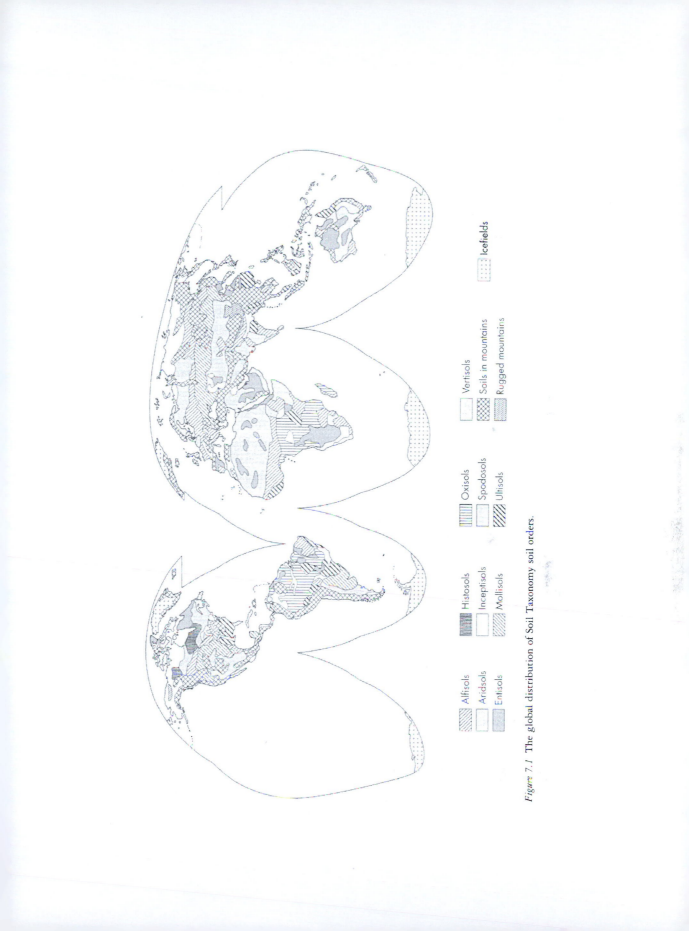

Figure 7.1 The global distribution of Soil Taxonomy soil orders.

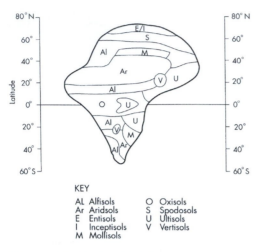

Figure 7.2 Distribution of Soil Taxonomy soil orders on
a hypothetical supercontinent.
Source: Based on Strahler and Strahler (1989)

essentially because of low temperatures. Transfers of
material by frost action and slope processes can be
significant. Mid-latitude soils are related generally
to precipitation amounts and seasonality. The
coolest and wettest areas are usually dominated by
spodosols. Mollisols occur in drier, more continental
areas with marked seasonality of precipitation,
alfisols in warmer and wetter areas, and aridisols in
the hottest and driest regions. Patterns in low-
latitude areas are usually determined by seasonality
in precipitation. Organic and mineral transformations are rapid in hot and humid areas, which means
that soils have limited organic horizons but often
deeply weathered profiles. These tend to be oxisols,
but where soils have been developing on old land
surfaces, ultisols are common. Soil patterns in
tropical wet/dry climatic zones are more variable.
They can be oxisols or ultisols, but alfisols are more
common, and vertisols occur given suitable base-
rich parent material. These broad relationships are
summarised in Table 7.1.

A similar sequence can be described using the
FAO classification system (FitzPatrick, 1983). In
northern latitudes, with low temperatures and

tundra vegetation and climate, gelic gleysols tend
to predominate. Where precipitation amounts are
slightly higher, humic gleysols and histosols are
found. Moving south, a gradual increase in precipitation and mean annual temperature brings in the
boreal zone of coniferous vegetation and podzols.
Gleysols may be found in waterlogged depressions.
In the next zone south, temperatures increase and
podzols are replaced by luvisols and podzoluvisols
under deciduous forest. The next major zone is that
of chernozems on the steppes, although there may
be a narrow zone of phaeozems in the transition zone
between deciduous forest and steppe. As the zone of
deserts is approached, soils change from haplic and
luvic kastanozems to xerosols. There is a gradual
decline in organic matter content and an increase
in the amounts of carbonates and soluble salts.
Depending on local conditions, saline and alkaline
soils (solonchaks, solonetz, solidic planosols) may be
found.

Soil zones to the south of northern deserts reflect
increasing amounts of precipitation in generally hot
climates. Rhodic cambisols and vertisols form in the
zone experiencing tropical wet/dry climates as well
as rhodic ferralsols. Xanthic ferralsols predominate
in tropical rainforest areas. These general patterns
are summarised in Table 7.2.

A slightly different interpretation of the global
pattern of soils has been provided by Paton *et al.*
(1995). They adopt a plate tectonics paradigm
and relate soil patterns to specific plate segments,
namely tensional margin, plate centre and compres-
sional margin. Determinative factors of soil formation will vary between plate segments (Table 7.3).
Plate centres dominate most continental landscapes,
especially Africa, Australia and Eurasia, and appear
to have the simplest soil patterns noted earlier.
Paton *et al.* (*ibid.*) relate these patterns to predominantly gentle topography with overwhelmingly
granite or granite-derived parent material and
a landscape that has been stable for millions of years.
Processes of soil formation would have been
expected to have reached their end-points. The
world soil map indicates that the most complex soil
patterns occur in areas that could be termed plate

Table 7.1 Relationships of Soil Survey Staff (1975) soil orders to broad bioclimatic zones

Bioclimatic zone	Soil orders	Annual precipitation range (mm)	Temperature patterns
Equatorial and tropical rainforest	oxisols, ultisols	1,800–4,000	Always warm (21–30°C)
Tropical seasonal forest and scrub	oxisols, ultisols, vertisols, some alfisols	1,300–2,000	Variable, always warm (>18°C)
Tropical savanna	alfisols, ultisols, oxisols	900–1,500	No cold weather limitations
Mid-latitude broadleaf and mixed forest	ultisols, some alfisols	750–1,500	Temperate, with cold season
Needle-leaf and montane forest	spodosols, histosols, inceptisols, alfisols	350–1,000	Short summer, cold winter
Temperate rainforest	spodosols, inceptisols	1,500–5,000	Mild summer and mild winter for latitude
Mediterranean shrubland	alfisols, mollisols	250–650	Hot, dry summers, cool winters
Mid-latitude grasslands	mollisols, aridisols	250–750	Temperate continental regimes
Warm desert and semi-desert	aridisols, entisols	<20	Highest temperatures on Earth
Cold desert and semi-desert	aridisols, entisols	20–250	Mean around 18°C
Arctic and alpine tundra	inceptisols, histosols entisols	150–800	Warmest month <10°C

Source: After Christopherson (1992)

margins, areas such as western North and South America, the Mediterranean region, and Southeast Asia. Thus, the Paton *et al.* (*ibid.*) synthesis is well worth considering, not least because plate tectonics appears to be providing a concept that integrates many Earth science disciplines.

Patterns on individual continents still appear to be dominated by broad climatic factors, but parent material, topography, and geological and geomorphological history are also important. This can be seen by examining soil patterns across Africa and Australia. The following account draws heavily on the excellent and exhaustive synthesis by Foth and Schafer (1980). Terminology is that of the Soil Survey Staff scheme.

Africa

Africa's surface, of about 30 million km^2, is 71 per cent soil-covered, with 20 per cent desert. The rest is bare rock, water and snow. Africa is a topographically simple landscape, being essentially a plateau sloping upwards from north to south. Lack of major mountain ranges ensures that a map of precipitation amounts shows a broad zonal pattern (Figure 7.3). Amounts are greatest at the equator, decreasing north and south to virtually zero in the desert areas. At the northern and southern extremes, precipitation increases again in the mediterranean climate of wet winters and dry summers.

Table 7.2 Relationships between FAO/UNESCO soil orders and climatic regimes

Type of climate	Soil orders
Wet equatorial	ferralsols[+++], acrisols[++], luvisols[++], nitosols[++]
Trade wind littoral	ferralsols[+++], acrisols[++], luvisols[++], nitosols[++]
Tropical steppe	xerosols[+++], vertisols[+++], arenosols[++], lithosols[++], solodic planosols[++], solonchaks[++], solonetz[++]
Tropical desert	lithosols[++], xerosols[++]
Tropical wet/dry	luvisols[+++], acrisols[++], ferralsols[++]
Humid subtropical	acrisols[+++], ferric cambisols[++], luvisols[++]
Maritime west coast	podzols[+++], cambisols[++], gleysols[++], histosols[++], podzoluvisols[++]
Mediterranean	luvisols[++], planosols[++]
Mid-latitude steppe	chernozems[++], kastanozems[++], phaeozems[++], solonchaks[++], solonetz[++], xerosols[++]
Mid-latitude desert	xerosols[+++]
Humid continental	gleysols[++], histosols[++], luvisols[++], podzols[++]
High-latitude	histosols[+++], podzols[+++], gleysols[++], lithosols[++], placic podzols[++]
Tundra	gelic gleysols[+++], histosols[++], lithosols[++]

[+++] abundant, [++] frequent
Source: Adapted from FitzPatrick (1983)

Table 7.3 Plate segments and characteristics

Environmental influence	Plate segments		
	Tensional margin	Plate centre	Compressional margin
Lithospheric material	Basaltic	Granite	Mixed
Topography	Steep with plateaux	Gentle	Steep
Vulcanism	Active non-explosive	None	Explosive
Seismicity	Localised	Weak	Strong regional

Source: After Paton *et al.* (1995)

In the equatorial zone, rainfall is generally 1,000–2,000 mm a year, with a decrease from west to east. Evaporation rates can exceed 8 mm a day, therefore soils may be dry even though the climate is generally humid. There is no true dry season in equatorial zones, but rainfall tends to be higher in March and October. Zones just north and south of the equator possess two short dry seasons. Further from the equator there is a distinct dry season, which gets longer as distance from the equator increases. Most of Africa experiences seasonal or periodic drought: the semi-arid areas have between 250 and 500 mm of precipitation, while deserts have less than 25 mm. Temperature is an important

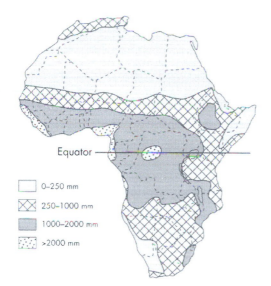

Figure 7.3 Precipitation map of Africa.
Source: After Foth and Schafer (1980)

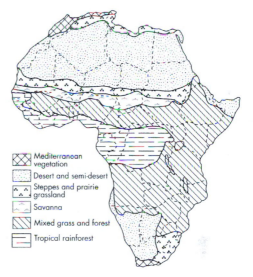

Figure 7.4 Vegetation map of Africa.
Source: After Foth and Schafer (1980)

factor. High temperatures in the tropics favour microbial activity, and rain water may be 15°C warmer than in temperate regions. At these temperatures, ionisation of water increases by a factor of 4, silica is eight times as soluble, and bases go into solution more readily. Although 5 per cent of Africa is covered by tropical rainforest, most is covered by desert or savanna vegetation (Figure 7.4). However, much of the vegetation has been affected by human activity in one way or another, as have some of the soils. Most rainforest vegetation is secondary growth, and overgrazing and soil erosion in savanna areas and the drylands are often intense (see Chapter 8).

Lack of extensive mountain ranges means that general climate–soil relationships are not interrupted by altitude–soil relationships (Figure 7.5). The desert areas (Sahara, Namib, Kalahari) are dominated by aridisols. Oxisols generally occur in equatorial regions, reflecting intense weathering under hot, humid climates on an old, level, unrejuvenated landscape. Chemical weathering is active throughout the year, and intense weathering

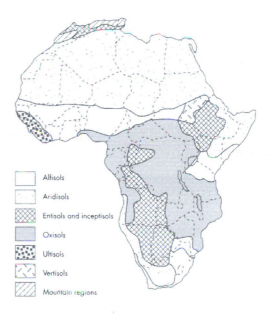

Figure 7.5 Generalised soil map of Africa.
Source: After Foth and Schafer (1980)

has resulted in the loss of silicates by hydrolysis. Plinthite frequently occurs. Alfisols are the third most extensive soil order and are generally located in the zone between oxisols and aridisols. In this zone, there is sufficient rainfall for appreciable soil development. Alfisols illustrate the problem with soil classification schemes as they occur in a variety of climates and are not necessarily typical of tropical areas. However, the same problem occurs with the FAO system with luvisols occurring in tropical wet/dry, humid subtropical, mediterranean and humid continental climates, although they can be differentiated into orthic, vertic, ferric, plinthic, etc. luvisols.

Entisols and inceptisols also cover large areas. Entisols are mostly psamments on sandy parent material. Inceptisols are usually aquepts on young parent materials on flat surfaces with poor drainage. They occur over large areas of the upper regions of the Nile and Zaire rivers. In North Africa, inceptisols are xerochrepts in the mediterranean climate. Ultisols are restricted to three areas, two along the equatorial west coast and a third area west of Lake Victoria. Africa has nearly 40 per cent of the world's vertisols, the largest area being in the upper Nile region on basic rocks rich in ferromagnesian minerals or on rocks with an appreciable calcareous content.

Australia

Australia also exhibits broad soil patterns that can be related to general climate. There are three main physiographic regions. Most of the west is a large flat shield area covering more than half the continent. It experiences low precipitation levels and contains very few permanent rivers. The eastern highlands form a relatively small coastal fringe. Between these two regions are the central lowlands, mostly below 150 m, which includes the dry salt flats associated with Lake Eyre and the Murray–Darling basin, which is the largest river system in Australia. Rainfall amounts are generally low, with 42 per cent of Australia classed as arid and only 9 per cent as humid; 40 per cent of the land receives less than 250 mm and 70 per cent less than 500 mm.

Soil patterns tend to reflect climate, but there are local factors that produce anomalies (Figure 7.6). Most of the soils in the central desert area are aridisols, but large areas of entisols also occur. Many aridisols show evidence of past climates, with plinthite gravels being relatively common. Some also show evidence of having been salty but are now acidic. Local conditions have given rise to five different saline or alkaline soils, some of which are unusual and possibly restricted to Australia. An area

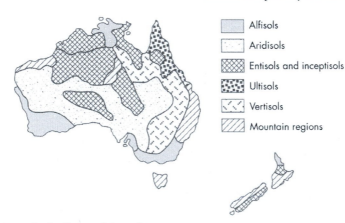

Legend:
- Alfisols
- Aridisols
- Entisols and inceptisols
- Ultisols
- Vertisols
- Mountain regions

Figure 7.6 Generalised soil map of Australia.
Source: After Foth and Schafer (1980)

of solodized brown soils occurs near Adelaide, containing large amounts of free carbonates and a calcic horizon below the argillic horizon. Calcium dominates the exchange complex, but soils are also high in magnesium. Much of the carbonate seems to have been wind-derived. The area was once covered in eucalyptus scrub called mallee. This vegetation seems unique to these soils, which is why they are known locally as mallee soils. Solonchaks and solonetz occur, as do solodized solonetz. Whereas solonetz have thin A horizons, solodized solonetz possess thick (30–45 cm) horizons. Both A and B horizons are acid, but the pH of the C horizon is often above 9. B horizons are strongly columnar. Solodised solonetz are generally believed to be highly weathered solonetz soils with intense leaching removing excess salts and leaving the exchange capacity dominated by magnesium and sodium. The fifth saline soil is the soloth, which is acid throughout with very strongly developed A1 and A2 horizons. There can be up to three times as much clay in the B horizon as in the A horizon, and the B horizon is strongly columnar to blocky. Hydrogen and magnesium dominate the exchange capacity, but amounts of sodium are still appreciable. The clay contrast is difficult to explain but is probably not simply the result of illuviation. The development of these soils is still a mystery.

Ultisols are found in the northeast of Australia, where the climate is warmer and more humid. Vertisols are important and occur on both base-rich rock and sediments derived from them. As with Africa, there is a great variation in alfisols from those on old northern landscapes in a monsoonal climate (ustalfs) to those formed in the mediterranean climate of the southwest corner (xeralfs) and along the southeast coast (udalfs). A more detailed account of soil patterns in Australia can be found in Paton *et al.* (1995).

REGIONAL SCALE

Soil patterns at the regional scale are often related to gross geological patterns, as the following two examples show. A typical soil landscape in the ridge-and-valley province of the Appalachians is shown in Figure 7.7. Ridges are usually composed of sandstone and valleys of limestone or shale. Soils on ridge tops are generally deep (>1 m), well drained, acid

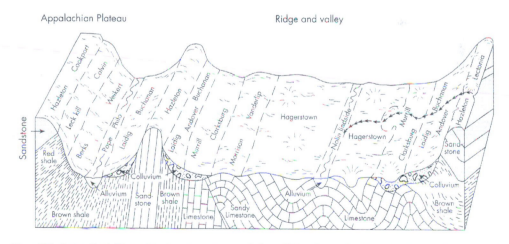

Figure 7.7 Generalised diagrammatic soil–landscape relations of the ridge-and-valley and adjacent plateau areas in central Pennsylvania.
Source: After Ciolkosz *et al.* (1986)

and sandy, with a high stone content. The commonest soil types are dystochrepts (Hazleton) and haplorthods (Lectonia). The Hazleton is the dominant soil on the ridge tops and is the most extensive soil in Pennsylvania. It also occurs on large areas of the unglaciated Appalachian Plateau. The sandy texture and cambic subsurface horizon are a function of the sandstone parent material, but the A and B horizons have been enriched in clay and silt, probably the result of aeolian inputs.

Soils on footslopes are developed in colluvium and are classified as fragiudults (Laidig, Buchanan), fragiaqults (Andover), fragiudalfs (Clarksburg) and hapludults (Murrill). They tend to be generally acid, well to poorly drained and medium- to fine-textured, with fragipans. The colluvium is highly variable and has been derived from sandstones on ridge tops, sandstone and shale on the main side slopes and shale and limestone from footslope areas. Soils developed on sandstone and shale colluvium are more poorly drained than those developed on limestone colluvium. Soils on valley floors differ according to whether the bedrock is shale or limestone. Soils on limestone are deep, well drained, acid and clayey hapludalfs (Hagerstown). Soils on shale are classified as dystochrepts (Berks, Weikert) and have silty textures. Thus the general patterns reflect the influence of parent material, but detailed differences also reflect climatically controlled processes that have acted with varying intensity over time in this area (Ciolkosz *et al.*, 1990: p. 258). This emphasises that a multi-factorial approach is often necessary when analysing soil patterns.

This is also true of the soil patterns developed in part of the North Downs in southeast England (Figure 7.8). The higher land is composed of chalk of varying characteristics and the valleys are usually excavated in clay. As with the Appalachian example, many of the soils have inherited characteristics, from previous climatically induced phases of development. The upland plateau areas are dominated by the Batcombe soil series, characterised by stagnogleyic palaeo-argillic brown earths (aquic paleudalf) with rubified clayey subsoil horizons containing

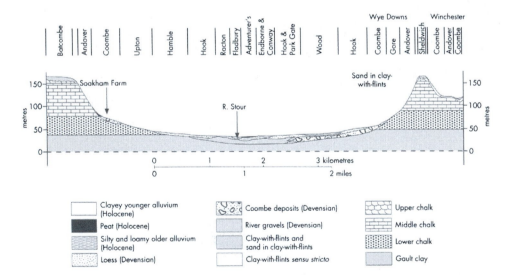

Figure 7.8 Section across the Stour valley, southeast England, showing the relationships of soil series to landforms and Quaternary deposits.
Source: From Catt (1986)

angular flint fragments from the chalk and small, round, flat pebbles from Tertiary formations. Soil development appears to have occurred through several interglacials, with cryoturbation during the cold Quaternary stages. Vertically oriented stones and polygonal patterns of narrow wedges can be found in the soils.

Some of the plateau areas possess well-drained palaeo-argillic brown earths (typic paleudulf) of the Sheldwich series. The steep valley slopes are covered with thin patches of frost-shattered chalk mixed with loess, and silty brown rendzinas (typic rendall) of the Andover series have formed. Material has been gradually removed from the slopes over thousands of years and deposited on the footslopes and in the river valleys. This has resulted in quite a complex pattern of soils. Fine silty calcareous brown earths (rendollic eutrochrept) of the Coombe series have formed on younger solifluction deposits and chalky Holocene colluvium. Typical brown earths (typic hapludalf) and palaeo-argillic brown earths (typic paleudulf) occur on other deposits on the gently sloping sides of broad, asymmetrical valleys.

Soils on the valley floors reflect the degree of waterlogging caused by high winter groundwater levels. These soils are typical argillic brown earths (typic hapludalf) of the Hamble series, gleyic and stagnogleyic argillic brown earths (aquic hapludalf) of the Hook series and typical argillic gley soil (aeric haplaquept) of the Park Gate series. Soils near the major rivers are developed on alluvium. These tend to be alluvial gley soils (typic fluvaquent) differing only in texture (fine loamy, Enborne series; fine silty, Conway series). Where the deposits are thinner and overlie solifluction deposits, a gleyic calcareous brown earth (fluvanquentic haplaquall) of the Wood series occurs. A groundwater gley soil (typic fluvaquent) of the Racton series occurs where thin alluvium overlies coarse gravels. Soils on younger alluvium are either clayey, non-calcareous alluvial gley soils (typic fluvaquent) of the Fladbury series or earthy eutro-amorphous peat soils (typic medihemist) of the Adventurer's series. At this medium landscape scale, the spatial variation of soils can be explained by the nature of Quaternary deposits.

LOCAL PATTERNS

At a more local scale, patterns of soils can still be recognised, but it is sometimes more difficult to explain them. Patterns can be recognised at four scales:

1 across the entire landscape;
2 relationships within smaller areas such as a single, reasonably sized drainage basin (fourth and fifth orders);
3 on individual slopes or along topographic transects (catenas);
4 related to individual landform units.

Landscape patterns

The nature of patterns *across an area* can be illustrated with respect to the granite area of Dartmoor, southwest England. Dartmoor is an upland plateau with an average elevation of about 400 m, rising to over 650 m in places. It receives over 2,500 mm of rainfall a year. Ridge tops slope from south to north and are separated by basin-like valleys and narrow gorges. Although it has never been glaciated, Dartmoor was subjected to intense periglacial activity during the cold phases of the Pleistocene Period. Six soil mapping units have been recognised (Hogan, 1982), and mapping these units produces a coherent pattern of soils (Figure 7.9). The terms are those used by the Soil Survey of England and Wales. Characteristic profiles are shown in Figure 7.10. The predominant map unit is the Hexworthy/Rough Tor, composed equally of iron pan stagnopodzols of the Hexworthy series and ferric stagnopodzols of the Rough Tor series. This unit is found on the lower plateau areas and on upper valley sides. Cambic stagnohumic gley soils of the Princetown series dominate the main upper plateau surfaces. Humic gley soils of the Laployd series are confined to basin, flush and valley bottom sites, and small areas of earthy oligo-amorphic peat soils of the Blackland series are found in small basin and flush sites. The lower moorland slopes are occupied by humic brown podzolic soils of the Moor Gate series,

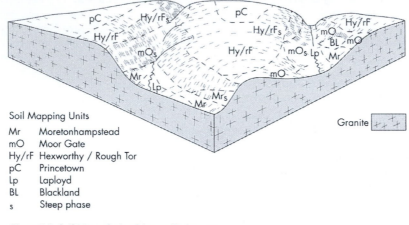

Soil Mapping Units

Mr	Moretonhampstead
mO	Moor Gate
Hy/rF	Hexworthy / Rough Tor
pC	Princetown
Lp	Laployd
BL	Blackland
s	Steep phase

Figure 7.9 Soil–site relationships on Dartmoor.
Source: After Harrod *et al.* (1976)

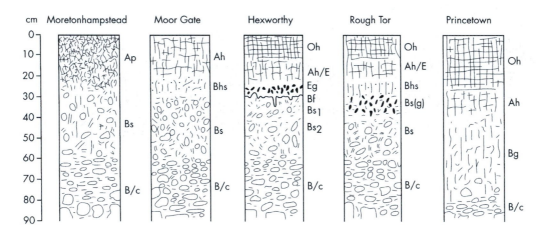

Figure 7.10 Major soil types on the Dartmoor granite.

and the brown podzolic soils of the Moretonhampstead series occupy the cultivated areas on the lower ground of the main valley sides and the granite margins. This broad spatial pattern is related to combinations of altitude, exposure, topography, drainage and, in some cases, human action.

Soil and landform patterns have been recognised by Tonkin and Basher (1990) in the eastern front range of the Southern Alps, New Zealand. In the dry Acheron valley, soils differed between the upper and lower valley. On southwest-facing lee slopes in the upper valleys, soils have developed in a thick mixture of loess and colluvium and possess either A/AB/Bw/Bc/C profiles or eroded derivatives. On northwest-facing slopes, a vertically striped pattern of soils has developed in scree with A/Ac/C profiles. Soil patterns in the more deeply incised lower valley are complex and reflect clearance by fire. A mosaic of A/C, A/Ac/C and A/Bw/C profiles occurs.

A rather similar mosaic of soils has been mapped by Burns and Tonkin (1982) in the southern Rocky Mountains. Soil patterns have been determined by altitude and aspect and the length of time snow lays on the ground (Figure 7.11). Soil terminology is that of the US Soil Survey Staff (1975). Extremely windblown sites (EWB) are found on the crests of the drainage divides, where soils (90 per cent dystric cryochrept, 10 per cent typic cryumbrept) are poorly developed and well drained with thin sandy A horizons over thin cambic B horizons. The windblown (WB) sites are found in a zone from the peaks to about 30 per cent downslope. Soils (80 per cent dystric cryochrept, 20 per cent typic cryumbrept) are similar to EWB soils but with thicker cambic B horizons. Minimal soil cover (MSC) sites occur in cols and on plateaux. Soils (80 per cent pergelic cryumbrept, 20 per cent dystric pergelic cryochrept) possess thick, fine-textured A horizons of aeolian origin overlying cambic B horizons. They tend to be the better-developed soils, and their A horizons have the highest organic matter contents. Early melting snowbank (EMS) sites occur on middle to lower, often gentle, slopes. EMS soils (60 per cent typic cryumbrept, 30 per cent pachic cryumbrept, 10 per cent dystric cryochrept) are fairly well drained, with the thickest A horizons. Late melting snowbank (LMS) sites are found on lower slopes, usually in leeward nivation hollows. Sites are rocky, with sparse vegetation, and soils (100 per cent dystric cryochrept) are poorly developed, moderately well-drained soils overlying weakly developed cambic B horizons. Semi-permanent snowbank (SPS) sites are found in nivation hollows with little vegetation cover. Soils (headwall, lithic cryothent; nivation hollow, pergelic cryoboralf or pergelic cryochrept) are poorly developed. Wet meadow (WM) sites occur below snowbank sites in depressions and on turf-banked terraces and lobes at the base of slopes. Soils are poorly drained, characterised by bog vegetation, with either A and/or O horizons of variable thickness overlying gleyed and mottled B and C horizons.

Drainage basin patterns

Individual, moderately sized *drainage basins* might be expected to show consistent soil patterns reflecting spatial landscape patterns. Arnett and Conacher (1973) have shown that as drainage basins develop, slope and landform units gradually become more integrated. With time, each slope and landform unit should develop a characteristic soil type. Soils in a fourth-order drainage basin on Dartmoor exhibit considerable variability, but patterns can be discernible (Gerrard, 1990). As a basic sampling unit and as a possible link with drainge basin development, slopes were defined according to the Strahler (1952) order of the stream at their base. Fourth-order slopes are dominated by iron pan podzols, but substantial variability exists over very short distances. The greatest variability is exhibited by the depth of the A–B horizon junction. Soils on third-order slopes are stonier and humic horizons are thinner than those on fourth-order slopes. Soils on second-order slopes are shallower and stonier, and the profiles often contain thin organic layers sandwiched between deposits of gravel.

Soil patterns within drainage basins occur because there are clear relationships between position, topography and surface processes. In a sixth-order

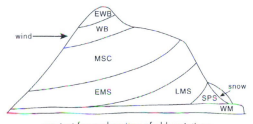

see text for explanation of abbreviations

Figure 7.11 Schematic representation of the synthetic alpine model.

Source: After Burns and Tonkin (1982)

drainage basin in Queensland, Australia, Arnett (1971) found that mean slope length and mean slope angle as well as mean maximum slope angle varied systematically with stream order. Similar results have been found in other studies (Figure 7.12). These relationships exist because stream order is associated with greater depth of incision as discharge of the basal stream increases. Slope length and relative relief will also increase. Arnett (*ibid.*) also found that the mean rate of slope convexity increased with stream order up to the fourth order and then declined. Many of the processes that determine the nature of slope form and angle also influence the development of soils. As there are systematic patterns of processes within drainage basins (Figure 7.13), there should also be patterns of soils.

Two-dimensional soil patterns can be examined by simple mapping, although the data can be

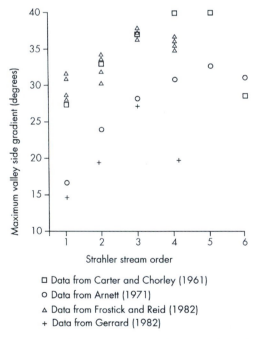

□ Data from Carter and Chorley (1961)
○ Data from Arnett (1971)
△ Data from Frostick and Reid (1982)
+ Data from Gerrard (1982)

Figure 7.12 Some relationships of mean valley side slope to stream order.

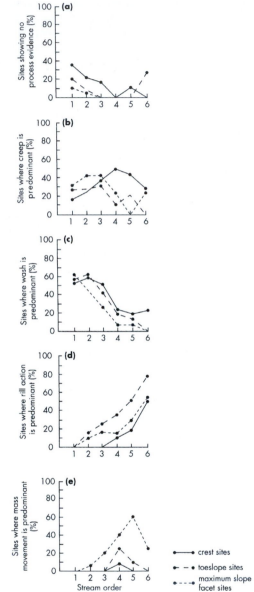

Figure 7.13 Variations in subaerial processes with slope component and stream order, Queensland, Australia. *Source*: From Arnett (1971)

manipulated in quite sophisticated ways. As an example, soil patterns in a small part of the Wyre Forest, England, will be considered. The Wyre Forest covers about 20 km² on Upper Carboniferous rocks to the west of the River Severn in the West Midlands of England. Bedrock lithology is an alternating sequence of shales, siltstones and sandstones. Although there have been phases of deforestation, the area has never been completely cleared of trees and is still covered with a dense mixed deciduous woodland. Initially, 6 km² was sampled, resulting in 201 sites spaced approximately 167 m apart. A non-hierarchical classification of soils was produced, which indicated six main clusters – the six groups that were identified in Figures 5.10 and 5.11. The characteristics of the groups and the equivalent soil series are shown in Table 7.4. Mapping by a technique termed Dirichlet tesselation produced a very fragmentary spatial distribution, which was only partially improved by applying spatial weighting (Figure 7.14). It was clear that a sampling distance of 167 m was too coarse to identify consistent patterns. Soil was then sampled more intensively and analysed in terms of variograms and mapped by a process known as kriging (Figure

7.15). This has identified relatively consistent patterns with spatial dimensions varying from 6 to 60 m. These patterns appear to reflect alternating rock outcrops of sandstone and shales.

There is some evidence to suggest that many of the soil patterns in the Wyre Forest reflect the influence of past periglacial activity and, in general, soil patterns on some slopes form an intricate mosaic, with soils of different ages existing side by side. Quite often, these soils are found to a greater extent as buried or fossil soils.

Slope–soil patterns (catenas)

There was an early realisation that patterns of soils occurred along topographic and slope transects and that these patterns were consistent and often repeated in a particular area. This led to the formulation of the concept of the catena. Milne (1935a: p. 197) produced probably the earliest definition of a catena as

a unit of mapping convenience, a grouping of soils which while they fall wide apart in a natural system of classification on account of fundamental and

Table 7.4 Summary characteristics of the six groups of the optimal classification of soils of the Wyre Forest

Group	Description	Soil series
1	Medium texture, typically sandy clay loam somewhat mottled and stony	Papworth
2	Clay loam to clay over strongly mottled grey clay	Dale variant
3	Reddish brown clay, increasingly clayey with depth, moderately mottled	Dale
4	Brown, sandy loam to loamy sand, somewhat stony, perceptibly podzolised	Rivington
5	Brown, sandy loam to clay loam becoming increasingly sandy with depth, moderately stony	Denbigh and lighter-textured series
6	Brown, sandy clay loam to clay loam, slightly mottled, shallow or very stony at depth	Denbigh and heavier-textured series

Source: After Oliver and Webster (1987a)

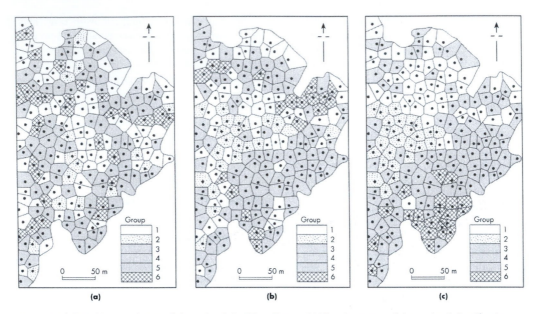

Figure 7.14 Dirichlet tesselation of the soils of the Wyre Forest. (a) The six groups of the optimal classification. (b) The groups created by a spatial weighting of 10. (c) The groups created by a spatial weighting of 17.5. *Source*: From Oliver and Webster (1987)

morphological differences, are yet linked in their occurrence by conditions of topography and are repeated in the same relationships to each other whenever the same conditions are met with.

Since this initial formulation, catenas have been recognised in a variety of areas and under a variety of climatic conditions.

The sequence of soils on a slope catena is a function of position, slope angle and drainage characteristics. As noted in Chapter 4, emphasis is usually placed on the difference between freely drained upper parts of slopes and imperfectly to poorly drained lower portions. Soil moisture conditions are related to slope angle as well as slope position. Thus there are sites where the influence of soil moisture is at a minimum and sites where the influence of soil moisture is at a maximum. Steep slopes reduce the amount of water infiltrating into the soil and may increase the removal of the upper parts of the soil through erosion. These processes

have their greatest influence where the ground surface slopes downwards continuously from the crest to the base of the slope. Thus it is probably incorrect to apply the term 'catena' to topographical transects that do not satisfy this requirement.

The result of these interactions is to produce a series of changes in soil properties from the upper to the lower members of the catena. In many catenas, especially those in tropical areas, the variation in soil colour is one of the more obvious signs of soil changes. Well-drained upland soils are usually reddish brown, indicating the presence of anhydrous iron oxide. The iron is quite well dispersed and is often partly attached to the clay fraction, making the clay appear red. Drainage is slower on the middle and lower parts of slopes, partly because of moisture seeping downslope from soils higher up the slope. Soils on lower slopes remain moist longer and dry out less completely, leading to an increasing degree of iron oxide hydration. The red colour changes to brown or yellow with increasing

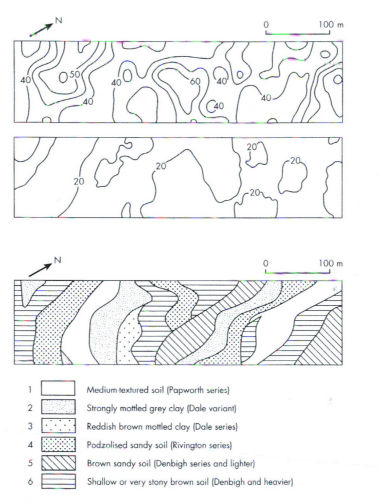

1 ▭ Medium-textured soil (Papworth series)

2 ▦ Strongly mottled grey clay (Dale variant)

3 ⣿ Reddish brown mottled clay (Dale series)

4 ▨ Podzolised sandy soil (Rivington series)

5 ◿ Brown sandy soil (Denbigh series and lighter)

6 ▤ Shallow or very stony brown soil (Denbigh and heavier)

Figure 7.15 Maps of percentage sand and percentage clay in the subsoil by kriging on a 20 × 25 m grid. Bottom: map of soil types on the two sides of a long transect.
Source: From Oliver and Webster (1987)

amounts of hydrated iron oxides such as goethite. Drainage is often very poor on the lowest slopes, and where the soil profile is waterlogged for appreciable periods, reduction of iron and other soil compounds occurs (see Chapter 3). Soil colour then changes to bluish grey, greenish grey or even neutral grey with mottles in the part of the profile where alternating aerobic and anaerobic conditions occur.

It is not only moisture status that integrates soils with topography and position. As noted in the previous section, the operation of surface processes is closely linked to soil processes. One of the most detailed syntheses of these interrelationships is provided in what is known as the 'hypothetical nine-unit land-surface model' (Dalrymple *et al.*, 1968; Conacher and Dalrymple, 1977). This is shown in

Figure 7.16. As this section is concerned with topographical transects, the two-dimensional version is used. However, the model has the advantage of being able to be used in three dimensions. It is a model based on form and contemporary geomorphological and pedological processes. It subdivides the slope profile yet integrates the components by considering material and water flow. Pedological processes and vertical water movement predominate on units 1 and 2. Mechanical and chemical eluviation within the soils is thought to be important. Surface soil creep becomes more important on the convex creep slope (unit 3), but soil processes are still important. Units 4 (fall face) and 5 (transportational midslope) are controlled by processes of weathering and mass movement. Geomorphological

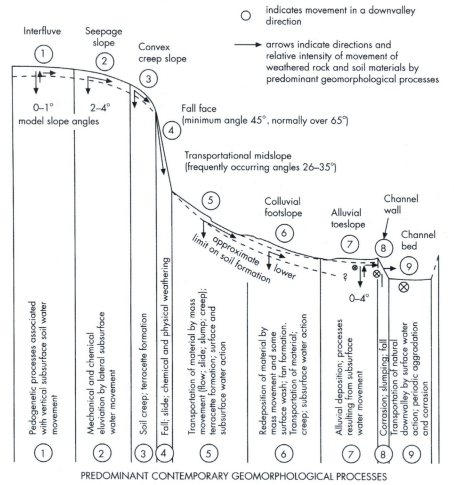

Figure 7.16 Hypothetical nine-unit land-surface model.
Source: From Dalrymple *et al.* (1968)

and pedological processes interact in the colluvial footslope (unit 6). The alluvial toeslope (unit 7) is controlled by subsurface water movement and occasional flooding of the river. Units 8 and 9 are fluvially controlled.

In some of these units, integration is produced by the mobilisation, translocation and redeposition of materials by overland flow. In other cases, the subsurface movement of water is more important. Unit 2 is defined as an area of the slope where the response to mechanical and chemical eluviation by downslope subsurface soil water movements makes it different to other parts of the landscape. Unit 5 is defined by the response to transportation of a large amount of soil material relative to other units (Conacher and Dalrymple, 1978).

These patterns can be integrated to produce wider drainage basin patterns by utilising stream order. The relative importance of the components in the nine-unit land-surface model and the sequence in which they occur on a slope will depend on the position of that slope within the drainage basin. Soil will also vary with these relationships. This results in soil–land-surface unit combinations within each valley or stream order (Table 7.5).

There is little doubt that the catena concept has been a useful one, but its adoption as a basic explanation of soil patterns on slopes has not been without problems. One of these concerns relationships on slopes where bedrock or parent material varies. The problem arises because the essence of a catena is the integration, downslope, between geomorphological and pedological processes. If parent material exerts a major influence on soil types, the patterns may be the result of this parent material influence and not the processes operating on the slope. Thus there has been an argument that catenas should be restricted to sequences on one-parent material, but this is difficult to apply, because parent material differences in the catena can occur even where the underlying geology is uniform. In the Dartmoor example noted earlier, the entire landscape is underlain by granite, but soils have developed on a variety of materials, initially derived from the granite. There is a variable thickness of

weathered granite, resting on unweathered bedrock with a layer of periglacially modified solifluction debris (head) on top of the weathered granite (Gerrard, 1989). Some soil patterns appear to be related to variations in these materials rather than to the underlying granite.

Milne was clearly aware of this situation, as the following quote shows:

> Since the first recognition of these catenary associations, it has become apparent that we have to deal with two classes of them. In one, the parent material does not vary, the topography having been modelled out of a single type of rock at both the higher and lower level. . . . In the other kind, the topography has been carved out of two superposed formations, so that the upper one is exposed further down the slope.
>
> (1935b: p. 346)

Catenas of the type suggested by Milne are common in West Africa, especially in northern Nigeria, where the landscape consists of sandstone- and ironstone-capped flat-topped hills with steep slopes rising above a gently undulating sandstone plain. The summits possess thin soils on the ironstone, the steep scarp slopes are covered with a shallow loamy soil over a sandstone and ironstone rubble, and the lower slopes and plains possess deep orange-brown to red sandy clays.

Soil patterns are more varied where slopes have formed on a variety of rock types and might then be simply related to rock type and not slope/soil processes. This tends to occur on longer slopes and over larger spatial scales, as exemplified by the Appalachian example noted earlier. It would seem to be incorrect to apply the catena concept to situations such as this, where soils are found to correspond more or less exactly with the underlying geological pattern.

In their specific detail all catenas are different, but there are sufficient similarities to enable groupings to be made on the basis of global climatic types. The greatest differentiation within catenas occurs in extreme situations such as those dominated by frigid or arid conditions. 'In all the rest of the world, under

Table 7.5 Relationships between soils, stream order and the units of the nine-unit land-surface model

stream order	Slope parameters					Soil types			Land surface units
	length (m)	maximum angle (deg)	mean angle (deg)	convexity (deg/100 m)	concavity (deg/100 m)	crest	centre	toe	unit combinations
1	132	16.7	9.4	16.4	37.0	deep red loam	deep red loam	deep red loam	1 : 5 : 2
2	254	24.1	16.9	30.0	41.0	red loam	red loam	acid red loam	1 : 5 : 3 : 2
3	273	27.4	19.0	37.0	33.0	shallow loam	skeletal loam	deep red podzol	1 : 5 : 3 : 5 : 8 : 9
4	351	31.4	20.0	45.0	24.0	skeletal loam	skeletal	deep red podzol	1 : 5 : 3 : 4 : 5 : 6 : 8 : 9
5	396	33.3	21.2	42.0	44.0	skeletal loam	skeletal	alluvial	1 : 5 : 3 : 4 : 5 : 6 : 7 : 8 : 9
6	476	30.9	16.0	40.0	51.0	skeletal loam	skeletal	alluvial	1 : 5 : 3 : 2 : 4 : 5 : 6 : 7 : 8 : 9

Source: After Arnett and Conacher (1973)

non-extreme conditions, the processes of slope erosion, slope deposition and pedogenesis are almost inextricably interwoven' (Ollier, 1976: p. 166). Much early work on catenas was undertaken in tropical savanna areas, and they have come to be regarded very much as the standard type, even though many variations occur (Table 7.6). In inselberg and pediment catenas, soils in a narrow belt surrounding the rocky inselbergs are shallow loamy sands with abundant coarse gravel. Incompletely weathered rock occurs near the surface. The most extensive catena member is to be found on the upper-middle and middle slopes. This is usually reddish in colour and of a sandy clay loam texture. Soils on low-middle to lower slopes are a variant of the middle slope member, being more affected by water seeping from upslope. The lowest slope members show evidence of a seasonally fluctuating water table. Catenas in tropical savanna areas without rock outcrops occur on convex–concave slopes. The crest is usually occupied by a dark red sandy clay with a well-developed structure, gradually changing into a yellowish red sandy clay with a well-developed structure where the slope angle steepens. The upper part of the concave slope section generally possesses a dark brown sandy clay loam

Table 7.6 Classification of tropical savanna catenas

---

1 Catenas with rock outcrops (inselberg and pediment)
   • with extensive pre-weathering
   • without extensive pre-weathering

2 Catenas with hard laterite
   • hard laterite as an upper slope feature
     (a) with massive laterite
     (b) with concretionary or detrital fragments only
   • hard laterite as a lower slope feature

3 Catenas without rock outcrops may be subdivided on the basis of underlying geology

---

Source: After Ollier (1959); Moss (1968)

overlying a mottled sandy clay. Where the slopes become gentler on the lowest parts of the concavity (1–2.5°), soils become sandy loam or loamy sand in texture, with the valley bottoms generally filled with black hydromorphic clay. Tropical rainforest catenas have been little studied but do not appear distinctive. A two-member catena can sometimes be identified consisting of a freely drained upper member and a poorly drained soil on the valley floor.

Arid and semi-arid areas constitute one of the extreme conditions stressed by Ollier (*ibid.*). Not only is the climate and lack of vegetation important, but the relief factor is often critical. The slope angle separating stable from unstable parts of the slope is more sharply defined in arid areas, and the position of the water table in relation to topography is crucial in determining the occurrence of saline and alkaline soils. The instability characteristic of most slopes in arid areas means that soils may be of varying ages. The steepest slopes possess the youngest soils, such as lithosols, whereas the more stable, lower-angled slopes possess the oldest soils. A simple pattern of young, relatively undeveloped soils on steeper slopes, followed by solonchaks on lower, moister slopes, is quite common.

The application of catenas to temperate environments is more difficult. There are patterns of soils on slopes, but they are not necessarily consistent. Geological diversity makes simple relationships uncommon. Also, many temperate areas are covered with a variety of superficial deposits, and past fluctuations of climate and vegetation types and the effect of human occupancy over the last 10,000 years have upset simple patterns between soils and topography. Many soil patterns represent a mixture of inherited and relict properties. It is instructive that the simplest catena-like patterns occur on small-scale depositional landforms such as glacial till and sand dunes.

The extreme conditions of cold tundra regions also produce relatively simple catena soil patterns. These usually reflect drainage conditions and the intensity of frost action in soils (Figure 7.17). The distinction between upland tundra and lowland or meadow tundra soils is usually on the basis of

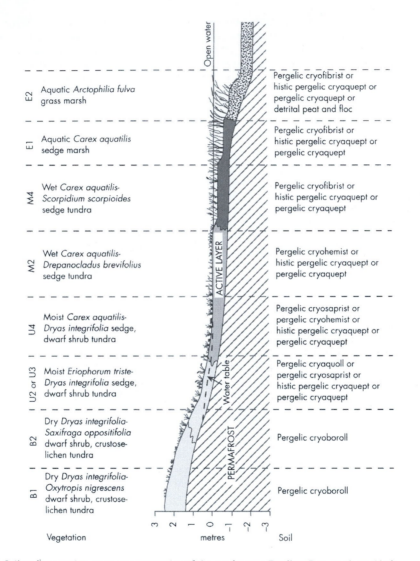

Figure 7.17 Soil and vegetation catena representative of the tundra near Prudhoe Bay, northern Alaska.
Source: After Walker and Everett (1991)

drainage. Upland tundra soils are drier, with a characteristic microrelief of non-sorted circles or earth hummocks. Lowland tundra soils are often underlain by ice wedges.

Soil and landform patterns

Many landscapes possess distinctive and often repetitive patterns of landforms. Such landforms might be erosional features but are more likely to be the result of deposition. Depositional landforms, by virtue of the way they have been formed, also tend to be composed of material with distinct characteristics. This combination of conspicuous form and distinctive material imparts a significant control on the soil types that develop. Thus landform patterns and soil patterns often coincide. Two examples, alluvial landforms and glacial depositional features, are chosen to illustrate this close relationship.

River-formed alluvial landscapes tend to be similar all over the world in terms of both landforms and processes. Soils may not be similar, but the patterns will be as soil materials and drainage conditions exhibit sharp boundaries related to major landform types. The main landform types that are likely to be found on most floodplains are shown in Figure 7.18. Each landform type is associated with a slightly different type of sediment and thus with a different soil. Natural levees, consisting of sandy material, occur adjacent to present and former river beds. Levees form the highest parts of the landscape and are more freely drained. Beyond the levees, lower-lying backswamp areas occur formed of heavier-textured and impermeable silts and clays. Accumulation of organic matter is common, and waterlogging is usually a characteristic. Levees are the most conspicuous fluvial landforms and can vary considerably in size. The Columbia River of North America has levees 50 m wide and 2.5 m high, but some levees are considerably bigger. The levees on the lower Saskatchewan River, Canada, are 1 km wide although only 4 m high (Smith, 1983). Levees are not always present. In much of the Missouri floodplain, levees are rare because of continual reworking by the river.

As levees and basins are universally important, it usually means that soil maps of widely differing rivers such as the Rhine, Mississippi, Indus, Tigris and Euphrates look very similar. In detail, the soils may differ because of climatic or large-scale geomorphological differences. This can be seen in the differences between alluvial soils of tropical and temperate areas. Floodplain soils in temperate areas are usually rich in mineral nutrients, partly due to former glacial action providing powdered fresh minerals to be transported and deposited by rivers.

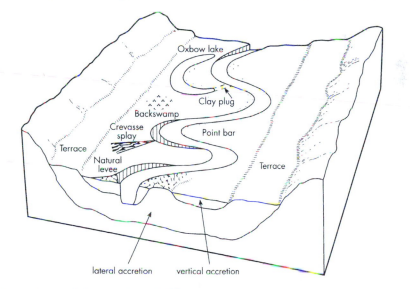

Figure 7.18 Landforms and deposits of a typical floodplain.

Many tropical and subtropical rivers have head-waters in deeply weathered igneous rock, and their deposits consist of quartz and other resistant minerals.

Crevasse splays form when an overbank flow cuts a channel through a levee and deposits material as a lobate sheet on the backswamp area. The depositional processes tend to produce a coarsening and then a fining upward sequence. The prograding front of the splay may bury waterlogged vegetal matter. A variety of bars can occur, but they are usually of three main types: longitudinal, transverse and point (lateral) bars. A succession of point bars and swales forms a meander scroll complex. This landform combination usually produces distinctive soil patterns, with the freely drained bars contrasting with poorly drained swales. Floodplains often contain a variety of lakes: oxbow lakes in abandoned meanders; embankment lakes in depressions dammed by levees; and serpentine lakes in sinuous abandoned channels. The laminated silts and clays of the lake deposits, high water tables and aquatic vegetation produce highly organic soils.

There are similar close relationships between soil and landform patterns in areas that have experienced glacial deposition. The type of deposit and landform and therefore the soil pattern will depend on the former relative positions of the ice. A model for a glacial depositional landscape (Figure 7.19) has been produced by Sugden and John (1976). These patterns are also imposed on soil patterns. In the zone of end moraines, the patterns tend to be parallel, reflecting the series of recessional moraines and interviewing depressions. In the zone of ice disintegration features, soil patterns form a mosaic reflecting the pattern of hummocks and depressions. Eskers impart a linear dimension to the patterns, and drumlins create a remarkably ellipical pattern. A linear dominance is re-established in the zone of drumlinised ridges.

A conspicuous pattern of swells and swales, associated with recessional moraines, has produced a banded pattern to the soils in many parts of North America. In Iowa, soils of the Clarion series occupy the swells and possess good natural drainage

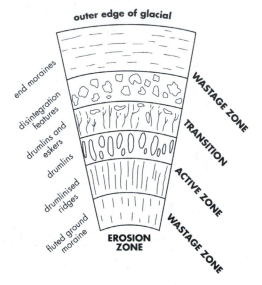

Figure 7.19 Glacial depositional landscape sequence.
Source: After Sugden and John (1976)

(Gwynne and Simonson, 1942). The A horizon (25–30 cm thick) is brownish black, with an intermediate texture, whereas the B horizon (10–50 cm) is dark yellowish brown. Soils of the swales have restricted drainage and belong to the Webster series. A horizons are thicker (30–50 cm) and are black with a heavy texture, while B horizons are dark grey and thinner (10–25 cm).

Some moraines can be large enough to impart catena-like patterns to the soils. A consistent series of soils has been recognised on the large Des Moines moraine in Iowa (Ruhe, 1969). The larger moraines in the Vale of York, England, possess similar soil relationships. Well-drained loamy brown earths of the Wheldrake series occupy the crest of the moraines. Downslope on the steeper sections of the moraines are brown earths of the Escrick series. These merge into acid brown earths (Kelfield series) on the lower slopes, followed by groundwater gleys (Fulford series) in the waterlogged depressions between moraines.

There is a glacial landscape composed of features variously described as 'knob and kettle', 'hummocky disintegration moraines' and 'ice disintegration features'. Landforms consist of a mixture of knolls or mounds and irregular depressions that have been formed by the stagnation and ablation of ice. Initially, mounds would possess a core of unablated buried ice. The way in which soils and landforms evolve is shown in Box 7.1. Such a landscape is characteristic of much of the Cheshire and Lanca-shire Plains in northern England (Crompton, 1966). Brown earths occur on the better-drained moderate to steep slopes of the mounds. They have A or Ap horizons of brown to reddish brown loamy sand with medium granular structure. Underlying horizons are strong brown to yellowish brown structureless sand. Layers of gravel and stones are common. Gleyed brown earths occur on the lower slopes, while in enclosed hollows, where major groundwater fluctuations occur, groundwater gley soils are found. Soils on lower slope portions are often silty and banded with organic matter.

Soils in the depressions are often arranged in concentric bands from the summit of crests to the basin centre. There are regular trends to soil prop-erties from hillock to depression (Walker, 1966). Organic matter increases as the thickness of the Ah horizon and total soil profile increase. Increased

Box 7.1

DEVELOPMENT OF SOILS IN HUMMOCKY DISINTEGRATION MORAINES

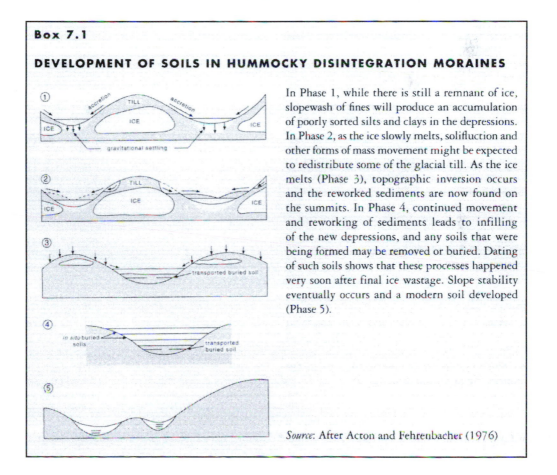

In Phase 1, while there is still a remnant of ice, slopewash of fines will produce an accumulation of poorly sorted silts and clays in the depressions. In Phase 2, as the ice slowly melts, solifluction and other forms of mass movement might be expected to redistribute some of the glacial till. As the ice melts (Phase 3), topographic inversion occurs and the reworked sediments are now found on the summits. In Phase 4, continued movement and reworking of sediments leads to infilling of the new depressions, and any soils that were being formed may be removed or buried. Dating of such soils shows that these processes happened very soon after final ice wastage. Slope stability eventually occurs and a modern soil developed (Phase 5).

Source: After Acton and Fehrenbacher (1976)

leaching and eluviation, the development of Ae horizons, lower pH values and lower base saturation occur in depression soils. In Iowa, soluble salts are lost from surface horizons and move vertically to underlying horizons and laterally to lower slope position. Prismatic structure increases towards the depressions as lime carbonates are removed from B horizons, with development of secondary blocky structure in illuvial profiles exhibiting Bt horizons.

SUMMARY

A spatial hierarchy of soil patterns can be recognised from the global scale to individual sites and landforms. The controls on those patterns vary according to the scale examined. At the global and continental scale, climate has always been regarded as the most important factor. But this may not be so, as there are distinctive global patterns of rock type and tectonic stability. Patterns of soils may be related to the major structural features of the Earth's surface. At the regional and local scale, parent material and geomorphological processes appear to be significant. This has led to the concept of soil geomorphology. An understanding of climate and environmental change is also necessary if soil patterns are to be explained.

ESSAY QUESTIONS

1 **Describe and explain the pattern of soils on a global scale.**

2 **What are the main problems in applying the soil catena concept to specific slopes?**

3 **What is the justification for using the term 'soil geomorphology'?**

FURTHER READING

Birkeland, P.W. (1984) *Soils and Geomorphology*, New York: Oxford University Press.

Daniels, R.B. and Hammer, R.D. (1992) *Soil Geomorphology*, New York: J. Wiley & Sons.

Foth, H.D. and Schafer, J.W. (1980) *Soil Geography and Land Use*, New York: J. Wiley & Sons.

Gerrard, A.J. (1981) *Soils and Landforms*, London: George Allen & Unwin.

Gerrard, A.J. (1992) *Soil Geomorphology*, London: Chapman & Hall.

Paton, T.R., Humphreys, G.S. and Mitchell, P.B. (1995) *Soils: A New Global View*, London: UCL Press.

Richards, K.S., Arnett, R.R. and Ellis, S. (eds) (1985) *Geomorphology and Soils*, London: George Allen & Unwin.

8

SOIL DEGRADATION

INTRODUCTION

There needs to be some justification for including a chapter on soil degradation in a general-purpose soils book. It may have been more appropriate to have titled the chapter 'Soil Use and Management'. However, such a chapter would inevitably have veered towards soil misuse and mismanagement. Thus it was thought more appropriate to examine the way in which the valuable resource, that is the soil body, is abused. Soil degradation seriously threatens agriculture and the natural environment. More than 97 per cent of the world's food comes from the land. Soil degradation, according to many authorities (e.g. Barrow, 1991; Blaikie and Brookfield, 1987), is now a major world environmental issue. Population and land-use pressures, especially the need to achieve greater agricultural productivity, are the main factors leading to degradation. Soil degradation affects from 30 to 50 per cent of the Earth's land surface. The scale of soil degradation worldwide has been examined by the Global Soils Degradation Data Base (GLASOD) project, executed by over 250 soil scientists from all over the world. This had the objective of 'strengthening the awareness of decision makers and policy makers on the dangers resulting from inappropriate land and soil management to the global well being, and leading to a basis for the establishment of priorities for action programmes' (Oldeman, 1988: p. 1).

Soil degradation can be defined in a variety of ways, but there is always the implication that it must be related, in some way, to human activity. Thus GLASOD defines soil degradation as 'human induced phenomena which lower the current and/or future capacity of the soil to support human life.' Factors causing soil degradation are shown diagrammatically in Figure 8.1. The emphasis is clearly on the effect of human activity on naturally occurring processes. Human-induced soil degradation processes can be divided into two categories (Box 8.1). The first is degradation by the displacement of soil, primarily by water erosion and wind erosion, which can cause degradation both on and off site. The second category concerns internal soil deterioration by physical and chemical processes and is confined to *in situ* effects. Although this is a useful subdivision, it is important to note that there will be interrelationships between the two groups. Soil compaction by agricultural machinery may lead to enhanced runoff rates and water erosion. Soil salinisation (see below) may lead to loss of soil structure and increased susceptibility to wind erosion.

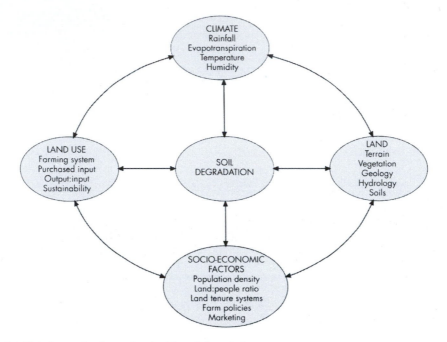

Figure 8.1 Main interacting factors involved in soil degradation.

Box 8.1

GLASOD SOIL DEGRADATION TYPES

Degradation type	*Description*
SOIL DISPLACEMENT	
Water erosion	
On-site	
Loss of topsoil	uniform loss by surface wash and sheet erosion.
Terrain deformation	irregular displacement characterised by major rills or gullies.
Off-site	
Sedimentation	of harbours, lakes, reservoirs.
Flooding	including by river bed filling, excessive siltation.

Box 8.1 continued

Degradation type	Description
Wind erosion	
On-site	
Loss of topsoil	uniform displacement by deflation.
Terrain deformation	uneven displacement characterised by major deflation hollows, hummocks or dunes.
Off-site	
Overblowing	encroachment on structures, roads and buildings, sand blasting of vegetation.
INTERNAL SOIL DETERIORATION	
Chemical deterioration	
Loss of nutrients	leading to reduced production and possibly soil acidification.
Pollution and acidification	excessive addition of chemicals (e.g. organic matter, acid rain, pesticides, metals).
Salinisation	caused by activities such as irrigation.
Discontinuation of flood-induced fertility	conservation methods that control flooding and lead to discontinuation of natural nutrient replenishment.
Gleyisation	as a result of waterlogging.
Other chemical problems	such as catclay formation in drained coastal swamps; negative chemical changes and development of toxicities in paddy fields.
Physical deterioration	
Sealing and crusting of topsoil	
Compaction	by heavy machinery on soil with weak structural stability, or on soils in which humus is depleted.
Deterioration of soil structure	due to dispersion of soil material by Na (and Mg) salts in the subsoil (sodification).
Waterlogging	human-induced soil hydromorphism; flooding and submergence (excluding paddy fields).
Aridification	human-induced changes of soil moisture towards an arid regime.
Subsidence of organic soils	by drainage, oxidation.

Source: Oldeman (1988)

There is thus a basic distinction between degraded and non-degraded soils. Non-degraded soils can be of two types, either 'stable' or 'non-used' by humans. Stable areas are either where human activity is minimal or where soil use is balanced by efficient soil improvement or protection procedures. The proportions of these categories worldwide are shown in Table 8.1.

The severity of soil degradation can be classified into a number of categories based on the degree of degradation and percentage of the area affected. The categories and their definitions are shown in Box 8.2. Table 8.2 shows the degree of soil degradation, continent by continent. Within continents, it is usually the susceptible dryland areas that have suffered most degradation (Middleton and Thomas, 1997) Worldwide, just over 20 per cent of susceptible dryland soils are degraded to some extent. A greater proportion of Europe's susceptible dryland soils suffer from degradation (32 per cent of the total area) than in any other continent. Most of these degraded soils occur in southern Portugal, Spain, Corsica, Sicily, southern Italy, Greece, Ukraine and Russia. In North America, only 11 per cent of susceptible dryland soils are degraded and mostly to a light or moderate degree. Asia, with 22 per cent of its susceptible dryland soils degraded, has a far greater proportion affected to a strong and extreme degree, but Africa, with 25 per cent of its susceptible drylands affected by soil degradation, has the highest percentage affected to a strong or extreme degree.

SOIL EROSION

Soil erosion by wind and water are natural processes that are often enhanced by human mismanagement of the land. Human-induced soil erosion, which has occurred ever since the natural vegetation cover has been disturbed, has been associated with most of the major civilisations around the world, examples being those that developed in the Mediterranean basin (e.g. Rubio, 1995) and pre-Spanish conquest Mexico (O'Hara et al., 1993). However, expansion and intensification of agriculture in the twentieth century has been the most significant. The 1930s 'Dust Bowl' on the Great Plains of the mid-west and southwest USA demonstrated the vulnerability of the soil to erosion. Large-scale ranching was developed from the 1860s onwards, and cattle numbers increased from 1 million in 1870 to 8 million in 1886. But it was the change to dryland farming that created the situation that culminated in the Dust Bowl (McGinnies and Laycock, 1988). Improved technology, especially tractors and combine harvesters, led to increased farm size and more tenant farmers. By 1930, 38 per cent of farms were run by tenant farmers who had litle stake in the long-term sustainability of the land. Droughts and dust storms are normal in this environment and occurred during the periods 1860–64, 1870–80 and 1910–18, but misuse of the land resulted in their increased severity.

The first large dust storm was recorded in the Texas Panhandle in 1932, moving east into

Table 8.1 Proportions of stable, soil-degraded and non-used land by continent (%)

Region	Stable land	Degraded land	Non-used land
Africa	39	17	44
Asia	48	18	34
Australia	61	12	27
Europe	61	23	16
North America	54	7	39
South America	72	14	14
World	52	15	33

Source: From Middleton and Thomas (1997)

Box 8.2

CLASSIFICATION OF SEVERITY OF SOIL DEGRADATION

None: There is no sign of present degradation from water or wind erosion, from chemical or physical deterioration; all original biotic functions are intact.

Light: The terrain is suitable for use in local farming systems, but with somewhat reduced agricultural productivity. Restoration to full productivity is possible by modifications of the management system. The original biotic functions are still largely intact.

Moderate: The terrain is still suitable for use in local farming systems, but with greatly reduced agricultural productivity. Major improvements are required to restore productivity. The original biotic functions are partially destroyed.

Strong: The terrain is not reclaimable at the farm level. Major engineering works are required for terrain restoration. The original biotic functions are largely destroyed.

Extreme: The terrain is not reclaimable and is impossible to restore. The original biotic functions are completely destroyed.

Source: After Middleton and Thomas (1997)

Table 8.2 Soil degradation degree by continent (million ha)

Region	Light	Moderate	Strong	Extreme	Total degraded	Total non-degraded
Africa	173.7	191.8	123.5	5.2	494.2	2471.4
Asia	294.5	344.3	107.6	0.5	746.9	3508.1
Australasia	96.6	4.0	1.9	0.4	102.9	778.3
Europe	60.5	144.5	10.7	3.1	218.8	731.7
North America	18.9	112.5	26.8	0.0	158.2	2032.7
South America	104.8	113.5	25.1	0.0	243.4	1523.1

Source: GLASOD

Oklahoma and Kansas. A reporter from the *Washington Evening Star*, covering the Oklahoma dust storms of 1935, coined the term 'Dust Bowl'. At its height in the period 1935–36, Great Plains topsoil was deposited on Washington DC, New York city and ships 2,000 km out at sea. By 1936, human-induced wind erosion had affected over 25 million hectares of land (Lockeretz, 1978), and the United States Conservation Service estimated that by 1937, 43 per cent of the land at the heart of the Great Plains had suffered severe wind erosion damage. The Dust Bowl led to new policies and actions, such as the Taylor Grazing Act of 1934, and the Soil Conservation Service was established to tackle soil erosion and advise farmers on the best use of the land. Another case study, illustrating the

Box 8.3

SOIL EROSION IN INDIA

Areas subject to different forms of erosion (MOA, 1985):

Type of erosion	Area (Mha)
Water	113.3
Wind	38.7
Saline alkali soils	8.0
Waterlogged soils	6.0
Ravines and gullies	4.0
Shifting cultivation	4.3
Riverine and torrents	2.7
Total	177.0

Impacts

Loss of soil fertility: 5.37 to 8.4 million tonnes of plant nutrients are lost every year due to soil erosion. Removal of 2.5 cm of topsoil leads to a 14 per cent decrease in maize yields; loss of 7.5 cm produces a 33 per cent decrease in yield.

Loss of rooting depth: In Maharashtra state, over 70 per cent of the cultivated land has been affected by soil erosion (32 per cent so severe that cultivation is no longer possible).

Loss of water resources: Due to erosion and loss of organic matter, subsoil holds less water. Faster runoff leads to flooding, and the period of baseflow is reduced. Streams and springs dry up.

Sedimentation: The annual rate of sedimentation in reservoirs is between 40 and 2,000 per cent more than anticipated at the time of project design. Data for 21 reservoirs show that sediment inflow is about 200 per cent more than design inflow.

Floods: Siltation of streams and channels reduces depths and water-holding capacity and increases width. The number of gullies and torrents increases. Since 1953, on average floods have inundated 4.9 million hectares of land annually in Assam, Bihar, Uttar Pradesh and West Bengal.

Other impacts include those on power generation, inland water transport and inland fish breeding.

Source: Data obtained from Khoshoo and Tejwani (1993)

nature and effects of soil erosion, is shown in Box 8.3.

It is not only such extreme examples that are significant. In some regions of England and Wales, as much as 25 per cent of the land most sensitive to erosion has eroded (Evans, 1988). Rates of water-induced soil erosion in lowland Britain vary from less than $1\,t\,ha^{-1}\,yr^{-1}$ to extremely high rates of $20\,t\,ha^{-1}\,yr^{-1}$ (Arden-Clarke and Evans, 1993). A five-year survey was carried out by the Soil Survey of England and Wales (SSEW) between 1982 and 1986. In 1982, 297 fields at fifteen selected localities were visited, and 148 were found to be eroding at rates ranging from 0.1 to 47.8 t ha^{-1} yr^{-1} (SSEW, 1983). By 1986, 1,769 fields suffering erosion had been located (Evans and Cook, 1986). However, this is clearly a major underestimation of the scale of the problem, as Reed (1979; 1983), by 1983, had located over 1,000 eroding fields in the West Midlands alone.

Whether soil is eroded or not depends on the balance between the erosivity of the eroding agents and the resistance of the soil. Erosivity concerns forces that can free particles from the soil surface and is controlled by the energy of the eroding medium (Box 8.4). Erodibility refers to the susceptibility of a soil to removal by the eroding forces and is a function of specific soil and terrain characteristics (Box 8.4). There is interplay between these characteristics, such that a crusted soil may be more susceptible to erosion by running water but less susceptible to wind erosion.

Erosion occurs when these erosional forces, triggered by the weather, overcome the resistance of the soil and vegetation to soil movement. The threshold at which this occurs depends on the type and condition of the soil and the type and degree of development of the vegetation or agricultural crop. Under some circumstances, this threshold can be quite low. Oliver and Gerrard (1996) have observed overland flow occurring on gentle slopes in the West Midlands after heavy rainfall lasting only 10 minutes. Under such conditions, 1–2 mm h^{-1} of rain with a total of 10 mm can be erosive.

Water erosion

Soil erosion by water occurs in five main, though often interrelated, ways. The impact of raindrops may liberate particles from the soil surface by splash erosion. This may move particles downslope, but its main effect is to loosen particles, which are then transported downslope by other means. If the ground surface is smooth, water may move across

---

Box 8.4

MAIN FACTORS AFFECTING SOIL EROSIVITY AND ERODIBILITY

Erosivity

of wind
wind velocity
frequency of strong winds
duration of windy events
wind turbulence

of water
rainfall intensity
frequency of rainfall events
rainfall duration
raindrop impact velocity
raindrop size

Erodibility

soil factors
soil particle (aggregate) size and shape
organic material content
clay content
cohesiveness
infiltration capacity
moisture content
porosity and permeability

surface factors
vegetation cover density
vegetation height and density
leaf/stem fineness
slope angle and surface roughness
presence/absence of surface crusts
plant orientation (wind erosion)

the ground as a thin, continuous film or as a large number of small, interconnected rivulets. This is known as sheetwash or inter-rill erosion. The role of sheetwash has probably been underestimated in most assessments of water erosion. Concentration of overland flow in small channels produces rilling. The depth and turbulence of water in rills is greater than for sheetwash, and larger particles can be picked up and moved. Rills are small enough to be removed by ploughing and need not reappear. Rills may develop into more substantial channels, known as gullies, which are more difficult to control. The fifth way in which soil particles can be removed is through subsurface channels by piping (see Chapter 4). These criteria have been used to assess the degree of water erosion (Box 8.5).

It is possible to note some general characteristics of soils that may make them more or less susceptible to water erosion (Box 8.6). Such knowledge is often not specific enough to be used in a predictive sense, but there have been a number of attempts to produce equations enabling soil loss to be predicted under specific conditions. Musgrave (1947) was one of the first to produce such an equation for erosion by surface water:

$$E = (0.00527)IRS^{1.35}L^{0.35}P_{30}^{1.75}$$

where E = soil loss, mm per year
I = inherent erodibility of soil on a 10% and 22 m long slope, mm per year
R = a vegetal cover factor
S = degree of slope (%)
L = length of slope in metres
P_{30} = maximum 30-minute rainfall

Following the publication of this equation there have been many attempts to refine it and incorporate the most important factors governing soil erosion. This work led to the development of the most widely used soil loss prediction equation, known as the universal soil loss equation (USLE):

Box 8.5

GLASOD METHODOLOGY FOR ASSESSING DEGREE OF WATER EROSION

Slight: For soils with a rooting depth exceeding 50 cm, part of the topsoil has been removed. Shallow rills with a spacing of 20–50 m may be present. Thin soils have rills at least 50 m apart. Perennial/optimal vegetation cover at least 70 per cent in pastoral areas.

Moderate: All the topsoil will have been removed from deep soils. Rills may be present, less than 20 m apart. Gully development will have occurred at a spacing of 20–50 m. This soil will have lost part of the topsoil and be likely to have rills with a 20–50 m spacing. Perennial/optimal vegetation cover in pastoral areas reduced to 30–70 per cent.

Strong: All the topsoil and part of the subsoil will have been removed from areas of deep soils, with moderately deep gullies less than 20 m apart. All the topsoil will have been removed from areas of thin soils, exposing bedrock, weathered bedrock or a hard pan. Perennial/optimal vegetation cover in pastoral areas will be less that 30 per cent.

Extreme: In general, the land is unreclaimable and impossible to restore. Virtually no soil will be left.

Source: After Oldeman (1988)

Box 8.6

TYPES OF SOILS SUSCEPTIBLE TO, OR RESISTANT TO, SPECIFIC FORMS OF WATER EROSION

Characteristics	*Susceptible to (+); resistant to (−)*
Soils sensitive to loss of organic matter because of low clay content	+ sheetwash erosion
Soils with moderately high clay contents that shrink and swell	+ rill and gully erosion − sheetwash erosion
Sandy soils with high infiltration rates; sensitive to crusting	if crusted + sheetwash erosion (also + wind erosion)
Soils with high content of water-soluble salts, which enhance dispersion	+ gully erosion + piping
Soils with high stone and rock fragment content	− all erosion types through enhanced infiltration; or + erosion through funnelling of surface flow

Source: Modified from Imeson (1995)

$$A = RKLSCP$$

where A = soil loss, kg m^{-2} s^{-1}
R = rainfall erosivity factor
K = soil erodibility factor
L = slope length factor
S = slope gradient factor
C = cropping management factor
P = erosion control factor

A nomogram has been produced to enable the erodibility potential of any soil under a variety of conditions to be estimated (Figure 8.2). The USLE may be used to predict average annual soil loss from a field or slope with specific land-use conditions; to guide the selection of cropping and management systems and conservation practices for specific soils and slopes; to predict the change in soil loss that would result from a change in cropping or conservation practices on a specific field; to determine how conservation practices may be applied or altered to allow more intensive cultivation; to estimate soil losses from land-use areas other than agricultural; and to provide soil loss estimates for conservationists to use in determining conservation needs.

A number of other models have been developed, such as CREAMS (chemicals runoff and erosion arising from agricultural management systems) and SLEMSA (soil loss estimation for southern Africa), to cater for specific problems and specific environments.

Recognition of the main factors governing soil erosion has led to attempts to model the individual processes in greater detail. Attempts to quantify the topographical influence have already been discussed in Chapter 4. The main factors discussed here are the erosivity of the rainfall and the erodibility of the soil.

Rainfall erosivity

Two main processes are involved in soil erosion by surface water: detachability and transport. Particles have to be detached from the surface of the soil before they can be moved by surface water movement. Detachment is largely achieved by rain splash, although only a very small proportion of rainfall energy contributes to splash erosion. Critical rainfall properties are duration and intensity,

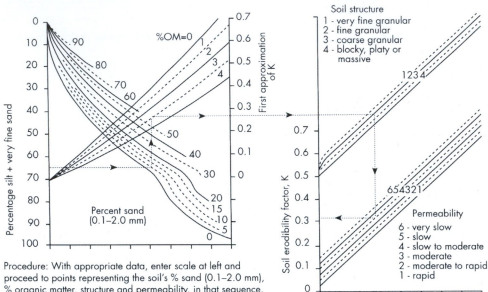

Procedure: With appropriate data, enter scale at left and proceed to points representing the soil's % sand (0.1–2.0 mm), % organic matter, structure and permeability, in that sequence. Interpolate between plotted curves. The dotted line illustrates procedure for a soil having: si + vfs 65%, sand 5%, OM 2.8% structure 2, permeability 4. Solution K = 0.31

Figure 8.2 Universal soil loss equation nomograph for calculating soil erodibility.

raindrop mass and size, and terminal velocity. These variables affect the kinetic energy and momentum of the rainfall. The median size of raindrops increases with rainfall intensity for low- to medium-intensity falls but declines sharply for high-intensity rainfall. Some of the more important attempts to relate factors such as kinetic energy to soil splash and to devise indices of rainfall erosivity are shown in Box 8.7.

Soil erodibility

Most studies have concentrated on the properties that influence soil detachability. Once a soil particle has been detached, its transportability is largely dependent on its size and shape as well as the energy and volume of the transporting medium. Usually less than 5 per cent of material eroded from slopes can be attributed to splash erosion alone (Morgan,

1977). However, detachment by splash is a vital necessity. Detachability depends on instrinsic soil properties such as humus and clay content, and extrinsic properties such as root density and vegetation cover. Most erodibility indices are based on properties affecting soil dispersion, soil aggregation, aggregate stability and the transmission of water. Aggregate (ped) size distribution is important because aggregates below a threshold size will be removed intact by flowing water. Aggregate size also influences water transmission. Aggregate stability is the ease with which large aggregates can be broken down by raindrop impact into smaller aggregates and become susceptible to erosion. Forces within aggregates leading to disruption are attributed to the swelling of oriented clay and to the compression of trapped air. Organic matter will strengthen the bonds within aggregates and may prevent disruption. Conaway and Strickling (1962) tested twenty-

Box 8.7

VARIOUS ATTEMPTS AT QUANTIFYING RAINFALL EROSIVITY

Bisal (1960) established that:

$$G = kDV^{1.4}$$

where G = weight of soil splashed (g)

k = a soil constant

D = raindrop diameter (mm)

V = impact velocity (m s^{-1})

Ellison (1945) produced:

$$E = kV^{4.33}D^{1.07}I^{0.65}$$

where k = a soil constant

V = raindrop velocity

D = raindrop diameter

I = rainfall intensity

Since 1945, there have been many attempts to devise the best index of rainfall intensity. Foster (1950), after a comparison of nine such indices, found that the best was:

$$index = I_{15}I_{30}(bn)^{1/3}$$

where I_{15} = maximum 15-minute intensity

I_{30} = maximum 30-minute intensity

b = function of the moisture content/infiltration relationship and storm intensity curve

n = number of peaks of intensity exceeding a given value

The rainfall factor in the universal soil loss equation is calculated by:

$$R = [\sum_{1}^{n}(1.2113 + 0.890\log I_j)(I_jT_j)]I_{30}$$

where R = rainfall erosivity factor

I_j = rainfall for a specific storm increment

T_j = duration of the specific storm

I_{30} = maximum 30-minute rainfall intensity

four indices of aggregate stability and found that the percentage weights of water-stable aggregates greater than 0.5 mm and greater than 2 mm in diameter were the most reliable measures. The most commonly used indices were also tested by Bryan (1969; 1977). These indices, as well as aggregation measures added by Bryan, are listed in Table 8.3. Middleton's dispersion ratio is the ratio of the amount of silt/clay in an undispersed soil sample with that in a sample treated with a dispersing agent. The erosion ratio is the ratio of the dispersion ratio to the colloid content/moisture equivalent ratio. The colloid content/moisture equivalent ratio is used as an index of water transmission, but as colloid content and moisture equivalent are closely related it is difficult to see the advantage of using both. The clay ratio is the ratio between sand and silt/clay; it attaches great importance to clay as a binding agent and ignores the influence of organic matter. A similar point can be made about the surface aggregation ratio, which is the ratio between the total surface area of particles larger than 0.05 mm and the quantity of aggregated silt/clay.

Interesting insights into some of these factors have been provided by a study of aggregates of forest and arable soils (Imeson and Jungerius, 1976): arable soil aggregates broke down very easily compared with forest ones (Table 8.4). This response

Table 8.3 Most commonly used soil erodibility indices

Dispersion ratio (Middleton, 1930)
Erosion ratio (Middleton, 1930)
Clay ratio (Bouyoucos, 1935)
Surface aggregation ratio (Anderson, 1954)
Water-stable aggregates (WSA): >3 mm, >2 mm, >1 mm, >0.5 mm
Slaking loss
Slaking loss/WSA: >2 mm
Moisture equivalent: >2 mm
Dry aggregates: >3 mm, 2–3 mm

Source: After Bryan (1977)

seems to be related to the contrasting nature of forest and arable soil aggregates (Table 8.5). Biological activity within forest soils produces a larger number of primary aggregates and much root tissue, which increases the aggregate stability. The higher organic content of forest soils will also make them more resistant to erosion. The greater amount of oriented clay in the arable soils makes them more susceptible to breakdown and therefore erosion.

This information suggests that it is changes brought about by agriculture that increase a soil's susceptibility to erosion. This can be illustrated by a specific example in the West Midlands of England (Oliver and Gerrard, 1996). The principal soil series, the Bridgnorth series, is a freely drained reddish brown soil with a pH of approximately 6–6.5. In texture, it is a sandy to coarse loamy sand with 80–90 per cent sand. The silt and clay fractions account for only 10–20 per cent of the total, with the clay content averaging less than 5 per cent. This texture means that the soil has a limited ability to retain water, and under desiccating conditions the surface is invariably dry and friable, making it susceptible to erosion, in particular by wind. The soil has a weak granular to subangular blocky structure but is sometimes apedal. There is in-

sufficient clay content to bind the coarser particles together, and poor aggregation makes the soil susceptible to slaking or structural collapse. Slaking produces a surface crust, which impedes infiltration of water into the soil. The soil organic matter content is low, about 1–2.5 per cent, and the coarse texture encourages the rapid breakdown of organic residues because of the oxidising environment. Breakdown products are also rapidly leached through the soil. The soil is easily worked at all times of the year, and continuous arable cultivation has led to a widespread decline in organic matter content, even though substantial amounts of plant residue are ploughed back.

It is not surprising that, with these characteristics, both water and wind erosion occur in the English Midlands. A major wind erosion event occurred in the spring of 1983, when the topsoil was desiccated (Fullen, 1985a). In most years, water erosion produces extensive rill systems, gullies and alluvial fans. Some of the rills can be up to 30 cm deep and often start in tractor wheel ruts, where the soil has been compacted (see later in this chapter). In a series of experiments near Wolverhampton, Fullen (1985b) measured infiltration rates before and after traffic over the soil. Infiltration was

Table 8.4 Relative erodibility of 4–5 mm aggregates from different slope positions under forest and farmland

Slope unit	Aggregates surviving test	Reduction in weight (%)	Organic material (%)	Material (g) splashed from cup after 50 drops
Forest				
0–2°	90	22	7.8	0.13
2–5°	50	56	10.2	0.15
5–11°	90	35	6.1	0.014
11°	75	35	8.6	0.62
Arable				
0–2°	5	96	4.0	0.36
2–5°	0	100	3.7	0.55
5–11°	5	95	4.0	0.83

Source: From Imeson and Jungerius (1976)

Table 8.5 Composition of aggregates under forest and farmland

	Slope class (forest)				Slope class (farmland)		
	0–2°	2–5°	5–11°	11°	0–2°	2–5°	5–11°
Zones of oriented clay	19	22	50	16	128	75	56
Humus particles	92	87	123	123	143	122	93
Living tissue	64	21	60	37	12	7	10
Rock fragments	11	20	7	14	20	20	12
Primary aggregates	39	50	37	55	11	10	19
Secondary aggregates	52	193	38	93	56	110	77

Source: From Imeson and Jungerius (1976)

reduced from $173 \, cm \, h^{-1}$ to $3 \, cm \, h^{-1}$. The preparation of a fine seed bed also means that the surface provides no resistance to the movement of water or wind over the surface.

Wind erosion

Wind erosion is an important process in many parts of the world and can be disastrous for agricultural productivity. It truncates the topsoil, removing the finer particles including organic matter, which will affect water-holding and infiltration properties and might lead to surface water erosion. In other areas, redeposited material buries soil and vegetation. The important wind variables are velocity, frequency, magnitude and duration. Surface factors include vegetation characteristics such as height, structure and density of plant cover, surface roughness and soil moisture status. Important soil variables are particle and aggregate size and cohesiveness, aggregate distribution and cohesiveness, essentially the same as those important in surface water erosion. The size and number of non-erodible particles play an increasingly important role as wind erosion progresses. There comes a point where non-erodible particles shelter erodible material, and a wind-stable surface is created. This final stage can be defined by a critical barrier ratio, which is the ratio of the height of non-erodible surface projections to the distance between projections that will barely prevent the movement of particles by the wind. Soil

deflation removes the finest particles in suspension, fine and medium-sized sand grains are moved forwards in a hopping motion by saltation, and the coarsest particles are moved close to the surface by creep. A methodology for assessing wind erosion is shown in Box 8.8.

An equation similar to the universal soil loss equation has also been produced:

$$E = IKCLV$$

where
E	=	annual erosion
I	=	soil and slope erodibility index
K	=	soil roughness factor
C	=	local wind erosion factor
L	=	length along prevailing wind direction
V	=	vegetal cover

Because of the complexity of the variables, each factor has to be treated separately, and computer programs are often required. One example for the local wind factor has been provided by Yaalon and Ganor (1966):

$$C = v^3/(P - E)^2$$

where v is average annual wind velocity (mile h^{-1}) at a standard height of 10 feet, and (P − E) is Thornthwaite's (1931) measure of precipitation effectiveness. Using this equation, they were able to define wind erosion zones in Israel.

Box 8.8

GLASOD METHODOLOGY FOR ASSESSING WIND EROSION

The methodology for assessing wind erosion recognises that there are uniform losses of topsoil, irregular or uneven displacement of soil material resulting in hollows, hummocks and dunes, and overblowing of material affecting physical structures such as roads and buildings.

Slight: In deep soils, topsoil partly removed and/or few (10–40 per cent of area) shallow (0–5 cm) hollows; in shallow soils, very few (<10 per cent) shallow hollows; in pastoral areas, ground cover of perennials of the original/optimal vegetation is >70 per cent.

Moderate: In deep soils, all topsoil removed or with common (40–70 per cent of area) shallow (0–5 cm) hollows, or few (10–40 per cent) moderately deep (5–15 cm) hollows; in shallow soils, topsoil partly removed or few (10–40 per cent) shallow (0–5 cm) hollows; in pastoral areas, ground cover of perennials of the original/optimal vegetation is 30–70 per cent.

Severe: In deep soil, all topsoil and part of the subsoil removed or with many (>70 per cent of area) shallow (0–5 cm) or common (40–70 per cent) moderately deep (5–15 cm) or few (10–40 per cent) deep (15 cm) hollows/blowouts; in shallow soils, all topsoil removed with bedrock or hard pan exposed; in pastoral areas, ground cover of perennials of the original/optimal vegetation is <30 per cent.

Source: After Middleton and Thomas (1997)

CHEMICAL DETERIORATION

Nutrient depletion

Soil nutrient depletion is a problem facing many parts of the world. Nutrients become depleted when their removal from the soil exceeds their addition. Losses occur through extraction by crops, by leaching and by erosion, and by volatilisation and denitrification. Burning of vegetation and stubble also leads to a nutrient loss. Inputs include application of fertilisers and manure, nitrogen fixation, atmospheric deposition in rain and dust, and enrichment by weathering of soil minerals. Crops that are demanding of nutrients often lead to severe nutrient depletion, sometimes to acidifcation (see below). Nutrient depletion also occurs through clearance of natural vegetation. Nutrient depletion criteria are examined in Box 8.9.

Fertilisers

Fertiliser addition is widespread and can lead to the fertiliser being leached to drainage waters, causing river and groundwater pollution. Fertiliser not leached may affect soil properties. Nitrogen, phosphorus and, to a lesser extent, potassium are the three main nutrients that are widely applied to soils as inorganic fertilisers. Nitrogen is used more in temperate environments, and its application has directed attention to nitrogen transformations and the transport of inorganic nitrogen species such as NH_4^+ and NO_3^-. With highly weathered tropical soils, interest is focused more on phosphorus. Interest in potassium centres on its exchange on soil colloid complexes in order to evaluate the efficiency of potassium fertiliser retention and plant availability in the rhizosphere.

Box 8.9

GLASOD METHODOLOGY FOR ASSESSING DEGREE OF NUTRIENT DEPLETION FOR SOILS IN DRYLAND AREAS

The criteria used to assess the degree of degradation due to nutrient depletion are organic matter content, parent material and climatic conditions. Nutrient depletion is identified by a decline in organic matter, P and/or cation exchange capacity.

Slight: Cleared and cultivated grassland or savannas on poor soils in tropical regions. Formerly forested areas cleared and cultivated in tropical regions on soils with relatively rich parent materials.

Moderate: Formerly forested areas cleared and cultivated on soils with moderately rich parent materials, where subsequent annual cropping is not being sustained by adequate fertilisation.

Severe: Formerly forested areas cleared and cultivated on soils with inherently poor parent materials, with a low cation exchange capacity, where all above-ground biomass is removed during clearing and subsequent crop growth is poor or non-existent and cannot be improved by the addition of N fertiliser alone.

Extreme: Formerly forested areas cleared by removal of all above-ground biomass, on soils with inherently poor parent materials, where no crop growth occurs and forest regeneration is not possible.

Source: After Oldeman (1988); Middleton and Thomas (1997)

There are five possible fates for nutrients applied to soils in fertilisers. Some will be taken up by plants and indirectly by animals, and some will be fixed by adsorption and exchange in the soil, but a lot will be lost by leaching, volatilisation and runoff. Much attention has focused on leaching, especially nitrate leaching, as this determines how much is lost to groundwater and water courses. Nitrate leaching is influenced by most of the factors that influence leaching in general. Soil texture, especially percentage clay, affects water transmission and also the development of anaerobic conditions, which favour denitrification. Denitrification is only a major problem in fertilised, waterlogged soils such as rice paddies. Under these conditions, significant losses of nitrogen can occur by denitrification. In most fertilised soils, it is quick transmission of water by macropore bypassing flow that is most significant. Nitrate concentration during macropore leaching depends on the time elapsed since NO_3^- fertiliser application. For newly applied fertiliser, a number of studies (e.g. Smettem *et al.*, 1983; Barraclough *et al.*, 1983) report rapid loss of NO_3^-, resulting in higher concentrations of NO_3^- in leachate than expected. After a certain time, NO_3^- is relocated within soil aggregates and is protected from leaching when bypass flow occurs. Haigh and White (1986) have shown that the fine crumb structure of A horizons and agricultural topsoils allows thorough mixing of percolating water with the soil and efficient diffusion of NO_3^- into the soil solution. Water then drains quickly through B horizons via macropores and cracks. Drainage water interacts with a very small contact surface in the B horizons, and therefore NO_3^- concentration remains high.

The relation of nitrate leaching to rainstorm characteristics and soil water drainage in a sandy podzol in the Canadian Great Plains has been studied by Cameron *et al.* (1978). They examined migration of NO_3^-–N and NH_4^-–N from an

NH$_4$NO$_3$ (ammonium nitrate) fertiliser. The NH$_4^-$ – N was lost quickly by nitrification, especially in summer, which masked any small NO$_3^-$ leaching losses. The main leaching losses occurred in late autumn and early spring, with nitrate moving down the soil as a diffuse bulge rather than as a sharp plug.

The distribution of nitrate concentration is called a nitrate profile. When a pulse of nitrate is added to the soil surface, the boundary between the soil solution and the pulse is initially sharp. Addition of water displaces the bulge downwards. If no mixing between the pulse and soil solution occurs, the boundaries remain sharp as leaching occurs. In general, the solution moves more rapidly in large pores but is delayed in small pores, thus some nitrate moves ahead, causing spreading. Leaching with no mixing is called piston flow, and the mixing process is known as diffusion. The characteristic of the pulse as it moves through the soil reflects the size, shape and continuity of the pores (Figure 8.3). A large number of mixing models have been developed (e.g. Burns, 1974; Jury *et al.*, 1976; Addiscott and Wagenet, 1985; Addiscott *et al.*, 1991), but all are basically simple modifications of water transport models.

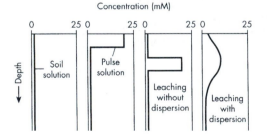

Figure 8.3 Movement of nitrate as a pulse.

Acidification

Some soils are moderately acidic as a result of naturally occurring factors such as long-term leaching and microbial respiration. Soil acidification is also caused by nitrification, which is the oxidative process of organic matter decomposition where NH$_4^+$ (ammonium) is converted to NO$_3^-$ (nitrate) ions by nitrifying bacteria, with H$^+$ ions as a by-product:

$$NH_4^+ + 1.5O_2 \rightarrow NO_3^- + 4H^+$$

Hydrogen ions are then able to displace base cations from the soil exchange complex, leading to soil acidification. Although this process occurs naturally, excessive use of inorganic nitrogen fertiliser can lead to soil acidification by this process. Excessive levels of NO$_3^-$ will encourage the leaching process and lead to acidification. In a series of experiments with different loadings of ammonium nitrate (NH$_4$NO$_3$), Chalmers (1985) was able to report significant decreases in soil pH with increased fertiliser loading. On one plot, pH was lowered from 6.9 to 5.4 after four years with an application of 750 kg ha^{-1} of nitrogen.

Soil acidification can also result from widespread needle-leaf afforestation (Hornung, 1985; Miller, 1985). Needle-leaf trees produce very acid litter, and their high canopy surface area allows the trees to 'scavenge' acid pollutants from the atmosphere, to release them later via throughfall and stemflow. Water transport through soils beneath needle-leaf trees is rapid, which means that reactions to the buffering processes are limited (Bache, 1983).

It is atmospheric deposition, either dry deposition or dissolved in rain water, that causes most concern (Fowler *et al.*, 1985). The main pollutants are sulphur dioxide (SO$_2$) and oxides of nitrogen (NO$_x$), which become dissolved in water as sulphuric acid or nitric acid. Values of pH from acid deposition in many urban and industrial areas are often less than 4.0. It is this type of activity that is 'scavenged' by needle-leaf trees, leading to marked increases in soil acidity (Hallbacken and Tamm, 1986).

Acidification leads to increased nutrient loss through leaching, especially of calcium and magnesium. Also, the mobility of aluminium increases as soils become more acid, especially when the pH drops below 5.0. Levels of aluminium are particularly high in the B horizons of some podzols

(Wilson, 1986). In soils where organic acids predominate, aluminium is mobilised as soluble organometallic complexes, whereas where mineral acids predominate, it is mobilised in its ionic, labile-monomeric form (Al^{3+}) (Adams et al., 1990). It is this form of aluminium which is toxic to many freshwater organisms (Ormerod et al., 1989).

The specific effect of acidification processes depends on the acid-neutralising or buffering capacity of the soil. Soils with base-rich, weatherable minerals have a high buffering capacity. Soils from acid parent rocks with appreciable quantities of resistant minerals such as quartz have a low buffering capacity. Therefore there tends to be a series of reinforcing processes whereby needle-leaf trees, which are usually planted on shallow, highly leached, acidic upland soils with a low buffering capacity, scavenge even more acidity and transfer it to the soil. Thus surface water acidification may be a problem in these areas. This was shown by Hornung et al. (1987) in a comparison of stream water under needle-leaf plantation and on open moorland in North Wales. Streams under forests were more acidic and contained higher levels of aluminium, sulphate and chloride. Liming is the most obvious way of combating such acidification, but it is curative rather than preventive. In Sweden, lakes are sometimes limed (e.g. Nyberg and Thornelof, 1988), but catchment liming is also practised (Brown, 1988; Dalziel et al., 1988).

Salinisation and sodification

Salinisation is the process by which salts accumulate in the soil, and sodification refers to the dominance of sodium ions within the soil exchange complex. These processes are primarily a problem in arid and semi-arid regions, where there is insufficient rain to leach away soluble salts, and upward soil water movement by capillary action plus evaporation leads to salt precipitation. Salinisation creates solonchak soils (see Chapter 6). Sodicity is a problem where parent materials and/or groundwaters are rich in sodium salts, producing sodic (solonetz) soils.

Soil salinity is thought to be the oldest soil pollution problem, and the failure of the Babylonian Empire is thought to have been partly the result of the accumulation of salts (Hillel, 1992). Human-induced salinisation and alkalisation are estimated to affect about 50 per cent of all irrigated land in arid and semi-arid regions (Abrol et al., 1988). The Aral Sea basin and the Euphrates valley are severely affected. It is estimated that about 45 per cent of the irrigated area of Syria is affected (Ilaiwi et al., 1992). In Iraq, over 60 per cent of the land irrigated under the 1953 Mussayeb Project had been affected by salinisation by 1970 (Iraq, 1980). Near Tashkent, on the Golodnaya Steppe, over 80 per cent of the irrigated area is saline, and irrigated areas at the delta of the Amu Darya, where it enters the Aral Sea, are severely affected (Smith, 1992).

Saline soils tend to occur in low-lying areas such as river floodplains and reclaimed estuaries and coastal areas, where water tables are high. In the latter case, there is the added problem of the presence of marine salts. Salts occur naturally in soils, dissolved in rain water and added as particulate input. According to Szabolcs (1976), salts occur because of one or more of the following:

1 Accumulation from poor-quality irrigation water.
2 Increase in the level of groundwater because:
   • the salt content of the groundwater accumulates in deeper soil layers;
   • the rising groundwater transports salts from deeper soil layers to the surface or surface layers;
   • the rising water table limits natural drainage and hinders the leaching of salts.
3 Lack or low effectiveness of drainage systems in irrigated fields.

The salt content depends on soil type as well as climate. Clay soils contain higher percentages of salts than sandy soils. Young, reworked sediments in upland river valleys contain the least percentages of salts. Groundwater of the major riverine plains is generally highly saline, but salinity values vary according to specific site characteristics. In the irrigated areas of the Murrumbidgee valley, Australia, salt has accumulated at the bases of stony or gravelly hillslopes as a result of downslope seepage. Salinisation can often be closely related to the type of river valley and the presence of buried former stream systems. In many places in the Loddon Riverine Plains, Australia, saturated channel sediments exist within 2–6 m of the surface, covered by more recent clay accretion. These sands have become waterlogged by seepage from swamps found outside the levees of a former river. Such alluvial sands may have either a detrimental or a beneficial effect on salt accumulation, depending on their ability to act as drains carrying salt away from the surface.

Topographical relationships of solodic soils are slightly different. Solodic soils are found usually on slopes just above floodplains and valley floors, where drainage potential is greater. On higher slopes, a leached and more acidic variety of solonetz soil, called a solod, is often found. Such soils were discussed in the context of Australia in Chapter 7. The catenary sequence of solonchak, solonetz and solod is widely distributed in the Indus valley of Pakistan, in the Nile valley of Egypt and Sudan, and in the Tigris and Euphrates valleys of Iraq and Syria.

Soil salinity is measured by the electrical conductivity of a saturated soil extract (EC_e). Saline soils are classed as having EC_e values of >4 mmho cm^{-1}. Sodicity is classified on the basis of exchangeable sodium percentage (ESP) or sodium adsorption ratio (SAR). Sodic soils usually have ESP values of >15 per cent. There is also a relationship between ESP, EC_e and pH. Soils are saline if the EC_e is greater than 4 mmho cm^{-1}, the pH is 8.5 or less and the ESP is less than 15 per cent. Non-saline sodic soils usually have a pH value greater than 8.5 and an ESP value above 15 per cent. Saline-sodic soils possess ESP values above 15 per cent and satisfy the salinity criteria.

Salinity has three main effects on plant growth. There may be direct toxicity from too much salt, high concentrations of sodium, chloride and boron being especially detrimental. High salt content also affects the ionic balance of plants and reduces the availability of water by lowering the osmotic potential. This last problem is sometimes called physiological drought. Sodicity presents even more problems as it directly affects soil structure. At ESP values between 10 and 15 per cent soil clays may swell and disperse, causing a breakdown in structure, especially when the soil solution is diluted by rain water or good-quality irrigation water. Smectite clays swell appreciably above ESP values of 15. The high alkalinity of sodic soils (pH up to 10.5) also causes organic matter dispersal, thus weakening soil structure.

The deterioration of structure has a number of consequences. Heavy-textured soils become more sticky and plastic when wet and hard when dry. Hydraulic conductivity is decreased, which may cause ponding on the surface. It then becomes more difficult to leach salts from the system. Also, if the soil surface becomes saturated during irrigation, air entry may be restricted, causing anaerobic conditions, denitrification and the production of plant toxins. Plants become more susceptible to salt damage in anaerobic soils. Clay dispersal due to high sodium contents leaves the soil surface more sensitive to the mechanical effects of rain such as capping, which will also reduce infiltration and cause waterlogging. This increased surface sensitivity will occur in soils with as little as 3–5 per cent exchangeable sodium.

Salinisation problems are not restricted to irrigated arable land but can occur in pastoral systems that become flooded for part of the year. In the Flooding Pampa of Argentina, salt concentration in the topsoil of grazed land was shown to increase considerably after each flooding event, whereas that in adjacent ungrazed land did not (Lavado and Taboada, 1987). This was attributed to greater compaction and therefore impeded drainage of the grazed land, producing less leaching and potentially more evaporation.

Reclamation of saline and sodic soils has to be managed carefully to suit the specific soil requirements. Soluble salts can be leached out of saline, non-sodic soils by using good-quality water. Saline-sodic soils require more careful treatment. The use of good-quality irrigation water will cause structural damage by removing soluble salts, allowing sodium ions to dominate the soil exchange complex. Water must have a low enough SAR to enable Ca^{2+} exchange for Na^+. It has been found that intermittent leaching is more effective than continuous ponding, as more time is allowed for salts to diffuse into the depleted outer regions of aggregates and then be removed during the next leaching cycle (Ellis and Mellor, 1995). Adding gypsum will help to maintain the soil solute concentration and supply calcium ions (Frenkel *et al.*, 1989; Armstrong and Tanton, 1992). It has been estimated that about 3 tonnes of gypsum per hectare is needed to reduce, by one unit, the ESP of a soil that is 15 cm deep (White, 1987).

Sodic non-saline soils are especially difficult to reclaim. The soil structure will already have been damaged and may be impossible to remedy. Surface application of gypsum will allow exchangeable calcium to be slowly added to the soil. Re-formation of soil structure will require grass or other forage crops, which may eventually improve soil structure in the topmost horizons, but the cost is high.

Pesticides

The vast majority of pesticides are organic chemicals and are usually used as insecticides, herbicides or fungicides, although specialised nematicides, miticides, rodenticides and molluscicides are available. As most pesticide molecules are uncharged and hydrophobic, they exhibit a stronger affinity for humus than other soil particles. However, paraquat behaves differently and adsorbs primarily on the negatively charged surfaces of clays.

Insecticides can be classified into three main groups. Organochlorine insecticides, which include DDT, lindane, dieldrin, aldrin and heptachlor, have the longest persistence in the soil of all organic pesticides. Their persistence is in the order DDT > dieldrin > lindane > heptachlor > aldrin, with a half-life of eleven years for DDT and four years for aldrin. DDT is now banned in many

countries along with some of the others, following exposure, in a series of publications (e.g. Carson, 1963; Mellanby, 1967) of their devastating ecological effects. Organophosphate insecticides also present problems as they are highly toxic to mammalian pets and dangerous to humans. They have shorter half-lives; six months for parathion, diazinon and demeton. Carbamate pesticides can kill a wide range and are used as molluscicides, fungicides and insecticides.

Five main groups of organic chemicals are used as herbicides. Phenoxyacetic acids, the best known of which are 2,4-D and 2,4,5-T, are selective herbicides and degrade quickly in the soil, but they have been banned in many countries because they have been implicated in growth abnormalities in animals. The toluidines and triazines are not easily leached from the soil, whereas phenylureas are less persistent as they are very soluble. Bipyridyls, of which paraquat is the most commonly used, are total herbicides, strongly adsorbed in the soil and very persistent.

A number of things can happen to pesticide compounds in the soil. They can be degraded by soil organisms or by physiochemical processes, adsorbed by soil organic matter, clay minerals and iron and aluminium sesquioxides, washed into water courses through leaching and runoff, or volatilised, resulting in atmospheric pollution. Biological degradation is concentrated in the surface soil and in the rhizosphere, where microbial numbers are highest. The degradation processes are also influenced by temperature, moisture, whether conditions are aerobic or anaerobic, pH, soil organic matter and clay mineral surfaces. After the initial application of pesticide, there is a lag time during which decomposer organisms become adapted to the particular pesticide. Organisms adapted to degrade one pesticide have the ability to break down compounds of similar molecular structure. This is known as co-metabolism. Non-biological degradation is achieved by hydrolysis, oxidation and reduction, and photo-decomposition. These are affected by the presence of clay mineral surfaces, soil organic matter and metal oxides. Adsorption on soil is a function of the chemical's surface charge (if any) and degree of aqueous solubility.

The main transport mechanisms affecting pesticides are leaching, surface runoff/erosion and volatilisation. Leaching potential depends on the amount of pesticide residue in the soil solution and the strength of binding to the soil, which is known as the pesticide's partition coefficient (McCall *et al.*, 1980). Macropore flow is an important process in deep soil contamination (Jury *et al.*, 1983). Pesticide leaching can be assessed by the scale known as the groundwater ubiquity score or GUS index (Gustafson, 1989). Pesticides likely to be leached have values of > 2.8, those unlikely to be leached values of < 1.8, pesticides with values of 1.8–2.8 being transitional. 2,4-D has a value of 1.65 and is unlikely to be leached, whereas atrazine (value 3.27) has a potential to move into groundwater.

Runoff potential depends on the nature and timing of rain events. Wauchope (1978) has described three different situations. Critical runoff events occur within two weeks of pesticide application and have at least 25 mm of rain and have a runoff volume that is ≥50 per cent of the rainfall input. Catastrophic runoff events are those that produce pesticide losses of ≥2 per cent of the applied amount, and transient runoff events are characterised by a small amount of rain, soon after pesticide application, but may produce very high pesticide concentrations in runoff due to a combination of low runoff volume and high amounts of pesticide residue in the soil. The last transport mechanism, volatilisation, is minimal in dry soils with low clay and organic matter contents.

Metals

Lead, zinc, cadmium, nickel and other metals are found naturally in soils whose parent materials are metalliferous. Thus soils in large areas of southwest England developed on metalliferous lodes contain appreciable quantities of a variety of minerals. However, what concerns us here are metals introduced into soils by a variety of human activities. In mining areas, metal concentrations in soils are increased many times during mining and smelting operations (Macklin, 1986). In urban areas, industrial pollution and exhaust gases will introduce metals such as lead into the atmosphere, which then reach the soil surface as either dry particulate matter or associated with precipitation. In 1981, the UK Department of the Environment conducted a survey of metal contamination in soils in Britain (Culbard *et al.*, 1988). A hundred household gardens were sampled in each of fifty-three cities, towns and villages. In most samples, total lead levels exceeded 200 mg kg^{-1}. Background lead levels are usually < 100 mg kg^{-1}. Lead levels were high in the inner areas of most large cities, with some exceptionally high levels (> 5,000 mg kg^{-1}) in some Derbyshire lead-mining villages.

Toxic metals can exist in soil in a number of forms. They may be adsorbed onto cations or be attached to clay and humus colloids and as organo-metallic chelates (Bridges, 1989). Two groups of toxic metals can be distinguished. Zinc, copper, nickel and boron have direct effects on crop growth if concentrations are high enough. Cadmium, lead, mercury, molybdenum, arsenic, selenium, chromium and fluorine, are not normally toxic to crops but may affect animals grazing on crops. Only boron is lost by leaching; all the others remain almost entirely as non-degradable contaminants. The availability of metals to plants depends mainly on CEC and pH values. Metals in soils with low CEC values are easily available to plants, whereas with high CEC values, metals are fixed in the soil through adsorption processes and are less available to plants. The mobility of metals is greater in soils with a pH of < 5.5 (Willet *et al.*, 1994). Recommended upper limits for major metals have been provided (Table 8.6). Dutch guidelines indicate when soils reach critical metal contents (Table 8.7). Metals are also introduced into soils via sewage sludge spread on fields. Metals usually present in sludges in the greatest amounts are zinc, copper and nickel. Little is known about the combined effect of different metals.

Table 8.6 Recommended upper limits of total metals in soils

Metal	Typical total metal content in uncontaminated soils		Recommended upper limit	
	(mg kg⁻¹)	*(kg ha⁻¹)**	*(mg kg⁻¹)*	*(kg ha⁻¹)**
Zinc	80	160	300	600
Copper	20	40	135	270
Nickel	25	50	75	150
Cadmium	0.5	1	3	6
Lead	50	100	250	500

* assumes 2,000 t ha⁻¹ to 15 cm depth
Source: After DoE/NWC (1981); ADAS (1987); Rowell (1994)

Table 8.7 Dutch guidelines for several soil metal pollutants

	Threshold concentration in soil (mg kg⁻¹ dry weight)		
	A	B	C
Chromium	100	250	800
Cobalt	20	50	300
Nickel	50	100	500
Copper	50	100	500
Zinc	200	500	3,000
Arsenic	20	30	50
Molybdenum	10	40	200
Cadmium	1	5	20
Tin	20	50	300
Barium	200	400	2,000
Mercury	0.5	2	10
Lead	50	150	600

A = reference (background) values; B = value indicative of further investigation; C = indicative value for cleaning operations
Source: After Thornton (1991)

Radionuclides

The 1986 Chernobyl nuclear accident in Ukraine has directed much recent attention to radionuclides in soils. The most common anthropogenic radionuclides found in soils are caesium (^{137}Cs, ^{134}Cs) and strontium (^{90}Sr). Their behaviour in soil depends on several soil characteristics, especially clay content and mineralogy, organic content, CEC, pH, ammonium content and nutrient status (Livens and Loveland, 1988). Radionuclides are immobilised most strongly in soils with high CEC and near neutral pH, when they are adsorbed onto clays, various mica varieties and humic materials. They are least well retained in acidic soils with low CEC and are therefore available for plant uptake. In the British Isles, the high-precipitation areas of northern and western Britain with acid soils were most affected by the Chernobyl fallout (Milne, 1987).

PHYSICAL DETERIORATION

Five forms of *in situ* physical soil deterioration have been recognised: compaction, sealing and crusting; dispersion of soil structure by sodification; waterlogging; aridification; and subsidence of organic soils. The nature of sodification has already been examined. The other four are now briefly described.

Compaction, sealing and crusting

Compaction usually occurs as a result of the repeated use of heavy machinery and trampling by animals on soils with low structural stabilities. The most common cause of sealing and crusting is clogging of soil pores by raindrop-dispersed fine-grained silt and clay particles. Crusting is the hardening

of the surface soil as it dries out. Crusts may also be the result of salinisation. As has been seen, this reduces infiltration rates and leads to surface runoff and probably soil erosion. It has been shown from measurements in Israel that sandy soil crusting can reduce the infiltration capacity from 100 to 8 mm h^{-1} and on loess soil from 45 to 5 mm h^{-1} (Morin *et al.*, 1981).

Compaction compresses the mass of soil into a smaller volume and therefore alters dry bulk density, porosity and resistance to penetration. Uncompacted soils usually have dry bulk densities ranging from 1.0 to 1.5 g cm^{-3}, while values for soils compacted by agricultural machinery usually exceed 1.5 g cm^{-3}. Values of 2.2 g cm^{-3} have been recorded along wheel ruts. The compacting forces will cause the realignment of clay particles in a direction parallel to the ground surface. This may produce a compaction or cultivation pan (Figure 8.4), which will tend to form 20–30 cm below the surface. A pan can also be created by ploughing continuously to the same depth. Soil can be compacted (poached) by grazing animals, especially during wet conditions in poorly drained or puddled areas.

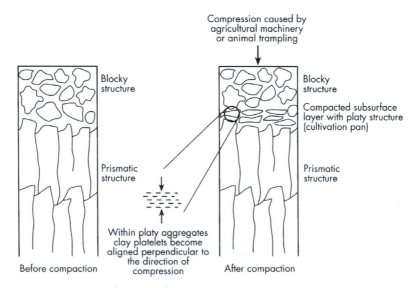

Figure 8.4 Soil compaction and the formation of cultivation pans.

A very similar process is known as hardsetting (Mullins *et al.*, 1987), but it occurs without the application of an external load. Soils prone to hardsetting have a low structural stability, aggregates collapse after wetting, and slaking leading to dispersal of clay and silt. They then dry hard without restructuring.

Waterlogging

The chemical effects of waterlogging have already been discussed. The physical effects are loss of air content, leading to plant stress and the inhibition or death of micro-organisms responsible for the biodegradation of organic material. Organic matter accumulates, leading to the release of acids and toxic substances.

Aridification

This is a human-induced diminution in soil water amounts. It may be caused by a lowering of the water table or by replacing natural vegetation with crops that demand greater moisture amounts for growth.

Subsidence of organic soils

The subsidence of peatland and organic soils is relatively common and is the result of three main effects: the compression of peat layers when the water table is lowered; shrinkage by desiccation; and the oxidation/mineralisation of organic matter.

SUMMARY

Soil, as a resource, is being increasingly abused throughout the world. It is being built upon, washed and blown away, and depleted of its natural chemical nutrients. Attempts to replace those nutrients have also led to changes in soil type and structure. Changing soil structure, especially the loss of organic matter, renders the soil susceptible to soil erosion. It appears that the lessons of the 1930s 'Dust Bowl' of the American mid-west have still not been learned. Soil degradation may be the biggest challenge facing mankind in the twenty-first century.

ESSAY QUESTIONS

1 **Discuss the nature and scale of soil degradation.**

2 **Examine the main controls on the nature and intensity of soil erosion.**

3 **Explain what is meant by the terms, salinization and sodification. How can the effects of salinization and sodification be reduced in soils?**

FURTHER READING

Ayoub, A.T. (ed.) (1991) *World Map of the Status of Human-induced Soil Degradation: A Brief Explanatory Note*, United Nations Environment Programme and International Soil Reference and Information Centre, Wageningen.

Barrow, C.J. (1991) *Land Degradation*, Cambridge: Cambridge University Press.

Bell, M. and Boardman, J. (eds) (1992) *Past and Present Soil Erosion: Archaeological and Geographical Perspectives*, Oxford: Oxbow.

Boardman, J., Foster, I.D.L. and Dearing, J.A. (eds) (1990) *Soil Erosion on Agricultural Land*, Chichester, J. Wiley & Sons.

Hudson, N.W. (1982) *Soil Conservation*, 2nd edn, Ithaca, NY: Cornell University Press.

McEwen, F.L. and Stephenson, G.R. (1979) *The Use and Significance of Pesticides in the Environment*, New York: J. Wiley & Sons.

Middleton, N. and Thomas, D. (1997) *World Atlas of Desertification*, 2nd edn, London: Edward Arnold.

Morgan, R.P.C. (1979) *Soil Erosion*, London: Longman.

Morgan, R.P.C. (1986) *Soil Erosion and Conservation*, Harlow: Longman.

Pimentel, D. (1993) *World Soil Erosion and Conservation*, Cambridge: Cambridge University Press.

GLOSSARY

abrasion pH A measure of the release of alkali cations when minerals are ground up in pure water.

acid A substance that releases hydrogen ions and a situation in which hydrogen ions exceed hydroxyl ions.

acidification An increase in acidity due to the removal of alkaline components.

actinomycetes The smallest organisms in the soil; they were originally thought to be fungi but are now known to be bacteria.

aerobic A situation where there is a continuous supply of molecular oxygen.

aggregate An amalgam of primary soil particles bound together by clay, humus, etc.

aggregate stability The resistance of aggregates to dispersal into their primary particles.

algae Simple unicellular or filamentous plants.

alkaline Conditions in which hydroxyl ions exceed hydrogen ions.

anaerobic A situation where molecular oxygen is absent.

anion An electrically charged particle that has gained an electron.

anion exchange capacity The total amount of anions that a soil can absorb by anion exchange.

argillan A depositional coating of clay on a soil mineral or particle.

atmosphere The layer of air surrounding the Earth, dominated by nitrogen and oxygen but also containing argon, water vapour, carbon dioxide and other gases.

atom The smallest unit in which an element can occur naturally.

Atterberg limits The boundary values of a soil between the liquid, plastic and solid states (see also **liquid limit** and **plastic limit**).

autotrophs Organisms capable of making their own food from inorganic materials.

azonal soils Poorly developed soils usually because of the effects of erosion and deposition or lack of time.

bacteria Micro-organisms, usually single-celled, that exist as free-living decomposers or parasites.

base A substance that reacts with hydrogen ions or releases hydroxyl ions.

base saturation This is defined as the percentage of base ions (non-hydrogen) that comprise the total exchangeable cations.

biomass The mass of living material in a community or specific area, usually expressed as a dry weight.

biosphere The layer which includes the fauna and flora that live above, at and below the Earth's surface.

biota A general term for all living organisms.

bioturbation The mechanical mixing of soil by biota.

bulk density The weight of soil per unit volume; can be wet bulk density, which includes water content, but is usually expressed on an oven-dry basis.

calcification The concentration of calcium compounds, particularly carbonate, by illuviation.

calcrete A hard, indurated deposit of calcium carbonate, also known as a petrocalcic horizon.

capillary water Water held by surface tension within the soil.

carbon cycle The movement of carbon through the various geospheres.

carbonation The reaction of minerals with carbonic acid.

catena A consistent downslope sequence of soils (see also **toposequence**).

cation A positively charged ion.

cation exchange The exchange between cations in solution and cations held on the surface of soil colloids.

cation exchange capacity The total amount of exchangeable cations that a soil can absorb, usually expressed as milliequivalents per kilogram.

chelation Formation of covalent bonds between metal atoms and organic molecules; also known as organic complexing.

cheluviation The process by which soluble organic complexes remove metal cations, especially iron and aluminium.

chronosequence The way in which specific soil types or soil properties vary with time.

clay A mineral particle <0.002 mm in diameter.

clay mineral A mineral possessing an orderly arrangement of atoms in layers.

clay ratio The ratio between sand and silt+clay.

clay–humus complex The combination of humus and clay with a complex chemistry.

colloid A negatively charged inorganic or organic particle of very small size, usually <0.002 mm in diameter.

consistency The behaviour of soil when it is handled, i.e. brittle, plastic, friable.

crevasse splay The spread of material deposited by a flow of water that cuts through a levee from the main river.

cryoturbation The mixing of soil brought about by repeated frost heaving and ice segregation.

cutan A coating, usually of fine material, on mineral grains (see also **argillan, skeletan**).

denitrification The rapid loss of nitrates by leaching.

denudational balance The balance between additions and removals from the soil body.

detachability The ease with which particles and aggregates of soil can be detached by processes of erosion.

dispersion ratio The ratio of the amount of silt/clay in an undispersed soil sample to that in a sample treated with a dispersing agent.

drumlin An elliptical landform composed of glacial deposits; often occurs in swarms aligned with the direction of ice flow.

duricrust A hardened crust at or near the surface.

earth hummocks Small (up to 1 m high) hummocks of soil and vegetation thought to be caused by the growth of segregated ice lenses within the soil.

ecosystem A group of organisms together with the physical environment with which they interact.

electron Small negatively charged particle forming part of the structure of atoms.

eluviation The general term for the transfer of soil material in solution, involving two main processes, **leaching** and **cheluviation**.

erodibility The susceptibility of a soil to erosion.

erosion ratio The ratio of the dispersion ratio to the colloid content/moisture equivalent ratio.

erosivity The strength of agents capable of producing soil erosion.

esker A long, sinuous ridge of fluvio-glacially deposited sands.

evaporation The diffusion of water vapour into the atmosphere from water exposed to the air.

evapotranspiration Evaporation plus the water lost to the atmosphere by plant transpiration.

exfoliation The weathering of rocks by the peeling away of surface layers, sometimes known as onion-skin weathering.

fabric The arrangement of solid particles.

field capacity The situation where the forces holding water to soil particles are equal to the downward forces of gravitational pull; the water content of a soil after drainage under gravity has been completed.

fine earth The fraction of the soil less than 0.002 mm in diameter.

fragipan A zone of high bulk density but which is not cemented.

geosphere The upper part of the Earth's crust in which pedological, biological and geomorphological processes operate.

gibbsite An aluminium oxide clay mineral, usually a weathering product.

gilgai Hummocky microrelief produced by large-scale expansion and contraction in soils rich in swelling clays.

gley A soil formed under poorly drained or waterlogged conditions, characterised by the reduction of iron to its ferrous state.

gravitational water The water that passes freely down the soil through the larger pores.

gypcrete A hard, indurated layer of gypsum-rich material; also known as a petrogypsic horizon.

haloclasty The weathering of rock by salt crystal growth.

head A term used in the British Isles for solifluction deposits.

heterotrophs Organisms that obtain nourishment from organic compounds.

horizon A distinct soil layer, approximately parallel with the soil surface, produced by specific soil-forming processes.

humification The process of humus formation.

humus The end-product of organic matter decomposition.

hydraulic conductivity The rate at which water can move through a soil.

hydrological cycle The movement of water, in all its forms, through the various geospheres.

hydrolysis The process whereby metal cations within mineral structures are replaced by hydrogen ions in soil water.

hydrosphere The zone that includes water in all its forms.

hygroscopic water Water held as microscopically thin films around individual soil particles.

illite A hydrous mica.

illuviation The precipitation or accumulation of material in soil after it has been leached from higher zones.

infiltration The process by which water enters the soil.

infiltration capacity The maximum rate at which water infiltrates the soil.

intrazonal soils Soils that appear to be related to parent material, topography or specific drainage characteristics.

ions Atoms, groups of atoms or compounds that are electrically charged (see also **anion** and **cation**).

isomorphic substitution The substitution of one ion for another within the mineral structure.

kaolinite An aluminosilicate clay mineral with a 1:1 crystal lattice form.

lateritisation The process whereby iron oxides are removed by eluviation from one part of the soil and deposited in lower soil horizons.

latosolisation The production of a material rich in oxides of iron and aluminium by the leaching of silica and bases from the soil.

leaching The movement of soil components in solution.

lessivage The movement of clays in colloidal suspension.

levee A river bank ridge formed by deposition when a river overtops its bank during a flood.

liquid limit The moisture content at which a soil changes from a plastic to a liquid state.

lithosphere The uppermost layer of the Earth's crust.

litter The dead components of biota or excrement added to the soil surface.

matric suction The combined effect of water adsorption forces and capillarity.

meander scroll complex A succession of point bars and swales on a river floodplain.

micelle A negatively charged colloidal particle that behaves like a giant anion.

moder Moderately decomposed matter in the surface horizon, with a pH of around 5.5.

montmorillonite An aluminosilicate clay mineral with a 2:1 expanding crystal lattice structure.

mor Organic matter in the surface horizon with limited decomposition and a pH of <5.5.

moraine A general term for material deposited by a glacier or ice sheet; can be classified by its position in relation to the glacier: e.g. terminal, lateral, medial, ground.

mull Well-decomposed organic matter in the surface horizon, with a pH of >5.5.

nitrification The biochemical oxidation of ammonium to nitrate.

nitrogen cycle The movement of nitrogen through the various geospheres.

nutrient A chemical element required by living organisms.

nutrient cycle The movement of nutrients within ecosystems.

overland flow The movement of water across the land surface; also known as surface runoff.

oxidation The removal of electrons from an atom, ion or a molecule.

pan A concentration of cemented plasma material in thin zones.

ped A comparatively permanent aggregate separated by planes of weakness.

pedogenesis The process of soil formation.

pedon The smallest three-dimensional unit of soil that can be analysed.

pedosphere The layer on the surface of the Earth in which soil-forming processes operate.

pedoturbation The mechanical mixing of soil.

percolation The downward vertical movement of water in soil.

permeability The ease with which gases or liquids pass through the soil.

pH An indication of soil acidity; a measure of the accumulation of hydrogen ions ($pH = -\log_{10} [H^+]$)

phytophage An organism that ingests plant material.

piping Subsurface pipes in soil created by a variety of natural processes.

plasma The components of soil that are capable of being moved, reorganised and concentrated.

plastic limit The moisture content at which a soil changes from a solid to a plastic state.

plasticity index The range of water content over which a soil is plastic.

plinthite A hard, indurated, iron-rich material formed by redeposition of iron in solution.

pore The space in the soil not occupied by solid material; also called void.

porosity A measure of the volume of pores to the total volume of soil.

protozoa Uni- or non-cellular organisms that live in water films; the smallest of soil animals ($5-40$ μm in length).

rain splash The impact of a raindrop on a bare soil surface, which might lead to the detachment of individual particles or aggregates.

regolith The superficial, usually weathered, mantle of rock debris on the Earth's surface.

rill A small channel created by concentrated overland flow; is small enough to be removed by ploughing.

ripening The changes that take place when waterlogged soils dry out.

salinity Conditions where there is a build-up of soluble salts because of insufficient water for the leaching process.

saltation The movement of soil particles in a series of leaps and jumps by wind or water.

sand A mineral particle between 2.00 mm and 0.05 or 0.06 mm in diameter, depending on the size classification adopted.

saprolite *In situ* weathered rock.

saprophage An organism that ingests dead organic matter.

saturation coefficient An assessment of the amount of water that a particular rock has absorbed when it is saturated.

sheetwash The movement of water across the ground as a thin, relatively continuous film; also known as inter-rill erosion.

silcrete A hard, indurated silica layer; also known as a duripan.

silicates The most abundant group of minerals in the soil, consisting of silica–oxygen tetrahedra linked in a variety of ways.

silt A mineral particle between 0.002 mm and 0.05 or 0.06 mm in diameter, depending on the size classification adopted.

skeletan A coating of sand/silt on mineral grains.

skeleton The relatively stable mineral grains and resistant organic components that comprise the main bulk of soil.

slaking The breakdown of soil aggregates in water.

smectite A group of swelling three-layer clay minerals, including montmorillonite.

sodicity Conditions where there are high concentrations of sodium.

soil association A group of taxonomic soil units occurring together over a region.

soil creep The very slow downslope movement soil without the aid of an erosion agent.

soil series A grouping of soil profiles characterised by a similar sequence of horizons developed on uniform parent materials.

solifluction The slow downslope movement of soil, usually saturated with water; usually somewhat quicker than soil creep; it is termed 'gelifluction' if frost action in the soil is involved.

solution The process whereby minerals break down as they dissociate into their component ions.

structure The size, shape and arrangement of particles, aggregates and voids.

sulphidisation The process of sulphur reduction, especially common in tropical estuarine and marine soils.

surface creep The dragging of particles and rocks across the ground surface by a flowing medium, i.e. along the bed of a river.

suspension The state in which particles are buoyed up by a fluid, either air or water.

swelling clays A group of clay minerals that increase considerably in volume due to water uptake.

texture The relative proportions of sand, silt and clay in a soil.

thermal conductivity The rate at which heat is transmitted through rock.

thermoclasty The mechanical break-up of rocks by heat; also known as insolation weathering.

throughflow The lateral, downslope movement of water through the soil.

toposequence A sequence of soils whose characteristics are related to surface form.

toposphere A term used to describe the surface of the Earth and the processes that operate there.

translocation The transport of material in soil, either in solution or in suspension.

transmission zone The zone through which water moves downwards into a dry soil behind the wetting front.

transportability The ease with which detached soil particles and aggregates can be moved by water or wind.

unloading The release of pressure on underlying rocks by erosion.

water table The interface between the capillary fringe and the saturation zone; will fluctuate seasonally in response to precipitation inputs. Perched water tables may exist locally above the main water table.

zonal soils Soils that appear to be related to broad bioclimatic zones.

zoophage An organism that is carnivorous.

REFERENCES

Abrol, I.P., Yadav, J.S.P. and Messoud, F.I. (1988) 'Salt-affected soils and their management', *FAO Soils Bulletin* 39.

Adams, W.A., Ali, A.Y. and Lewis, P.J. (1990) 'Release of cationic aluminium from acid soils into drainage water and relationships with land use', *Journal of Soil Science* 41: 255–68.

ADAS (1987) *The Use of Sewage Sludge on Agricultural Land*, Booklet 2409, MAFF Publications, Alnwick.

Addiscott, T.M. and Wagenet, R.J. (1985) 'Concepts of solute leaching in soils: a review of modelling approaches', *Journal of Soil Science* 36: 411–24.

Addiscott, T.M., Whitmore, A.P. and Powlson, D.S. (1991) *Farming, Fertilizers and the Nitrate Problem*, Wallingford: CAB International.

Anderson, D.H. and Hawkes, H.E. (1958) 'Relative solubility of the common elements in weathering of some schists and granite areas', *Geochimica et Cosmochimica Acta* 14: 204–11.

Anderson, H.A., Berrow, M.L., Farmer, V.C., Hepburn, A., Russell, J.D. and Walker, A.D. (1982) 'A reassessment of the podzol formation process', *Journal of Soil Science* 33: 125–36.

Anderson, H.W. (1954) 'Suspended sediment discharge as related to streamflow, topography, soil and land-use', *Transactions of the American Geophysical Union* 35: 268–81.

Arden-Clarke, C. and Evans, R. (1993) 'Soil erosion and conservation in the United Kingdom', in Pimentel, D. (ed.) *World Soil Erosion and Conservation*, Cambridge: Cambridge University Press, pp.193–215.

Arkley, R.J. (1963) 'Calculation of carbonate and water movement in soil from climatic data', *Soil Science* 96: 239–48.

Arkley, R.J. (1967) 'Climates of some great soil groups of the western United States', *Soil Science* 103: 389–400.

Armstrong, A.S.B. and Tanton, T.W. (1992) 'Gypsum applications to aggregated saline-sodic clay topsoils', *Journal of Soil Science* 43: 249–60.

Arnett, R.R. (1971) 'Slope form and geomorphological process: an Australian example', *Institute of British Geographers Special Publication* 3: 81–92.

Arnett, R.R. and Conacher, A.J. (1973) 'Drainage basin expansion and the nine unit landsurface model', *Australian Geographer* 12: 237–49.

Avery, B.W. (1973) 'Soil classification in the Soil Survey of England and Wales', *Journal of Soil Science* 24: 324–38.

Avery, B.W. (1980) *System of Soil Classification for England and Wales (Higher Categories)*, Soil Survey Technical Monograph No. 14, Harpenden.

Avery, B.W. (1990) *Soils of the British Isles*, Wallingford: CAB International.

Avery, B.W., Clayden, B. and Ragg, J.M. (1977) 'Identification of podzolic soils (spodosols) in upland Britain', *Soil Science* 123: 306–18.

Bache, B.W. (1983) 'The role of buffering in determining surface water composition', *Water Science and Technology* 15: 33–45.

Barraclough, D., Hyden, M.J. and Davies, G.P. (1983) 'Fate of fertilizer nitrogen applied to grassland: I. Field leaching results', *Journal of Soil Science* 34: 483–97.

Barron, E.J. (1986) 'Palaeoclimates and economic geology', Society of Economic Palaeontologists and Mineralogists, Short Course Notes No. 18, 162pp.

Barron, V. and Torrent, J. (1986) 'Use of the Kubelka–Monk theory to study the influence of iron oxides on soil colour', *Journal of Soil Science* 37: 499–510.

Barrow, C.J. (1991) *Land Degradation*, Cambridge: Cambridge University Press.

Bates, R.L. and Jackson, J.A. (1987) *Glossary of Geology*, 3rd edn, Alexandria, Va: American Geological Institute.

Bates, T.F. (1962) 'Halloysite and gibbsite formation in Hawaii', *Clays and Clay Minerals*, Ninth National Conference on Clays and Clay Minerals, Pergamon Press, pp.215–38.

Baxter, F.P. and Hole, F.D. (1967) 'Ant (*Formica cinerea*) pedoturbation in a prairie soil', *Proceedings of the Soil Science Society of America* 31: 425–8.

Bernier, N., Ponge, J.F. and André, J. (1993) 'Comparative study of soil organic layers in two bilberry–spruce forest stands (*Vaccinio piceetea*) in relation to forest dynamics', *Geoderma* 59: 89–108.

Birkeland, P.W. (1974) *Pedology, Weathering and Geomorphological Research*, New York: Oxford University Press.

Birkeland, P.W. (1984) *Soils and Geomorphology*, New York: Oxford University Press.

Birkeland, P.W. (1990) 'Soil-geomorphic analysis and chronosequences – a selective overview', *Geomorphology* 3: 207–24.

Bisal, F. (1960) 'The effect of raindrop size and impact velocity on sand splash', *Canadian Journal of Soil Science* 40: 242–5.

Bjerrum, L. (1967) 'Mechanism of progressive failure in slopes of overconsolidated plastic clay and shales', *ASCE Journal of Soil Mechanics Foundation Division* 93, SM5: 3–49.

Blackwelder, E. (1933) 'The insolation hypothesis of rock weathering', *American Journal of Science* 26: 97–113.

Blaikie, P.M. and Brookfield, H.C. (1987) *Land Degradation and Society*, London: Routledge.

Bockheim, J.G. (1980) 'Solution and use of chronofunctions in studying soil development', *Geoderma* 24: 71–85.

Bouma, J., Jongerius, A., Boersma, O., Jager, A. and Schoonberbreek, D. (1977) 'The function of different types of macropores during saturated flow through four swelling horizons', *Journal of the Soil Science Society of America* 41: 945–50.

Bouyoucos, G.J. (1935) 'The clay ratio as a criterion of susceptibility of soils to erosion', *Journal of the American Society of Agronomists* 27: 738–41.

Brewer, R., Crook, A.W. and Speight, J.A. (1970) 'Proposal for soil-stratigraphic units in the Australian stratigraphic code', *Journal of the Geological Society of Australia* 17: 103–11.

Bridges, E.M. (1989) 'Toxic metals in amenity soil', *Soil Use and Management* 5: 91–100.

Brown, D.J.A. (1988) 'The Loch Fleet and other catchment liming programs', *Water, Air and Soil Pollution* 41: 409–15.

Brunsden, D. (1979) 'Weathering', in Embleton, E. and Thornes, J. (eds) *Process in Geomorphology*, London: Edward Arnold.

Bryan, R.B. (1969) 'The relative erodibility of soils developed in the Peak District of Derbyshire', *Geografiska Annaler* 51: 149–59.

Bryan, R.B. (1977) 'Assessment of soil erodibility: new approaches and directions', in Toy, T.J. (ed.) *Erosion: Research Techniques, Erodibility and Sediment Delivery*, Geo Abstracts: Norwich.

Buckland, P.C., Gerrard, A.J., Larsen, G., Perry, D.W., Savory, D.R. and Sveinbjarnadottir, G. (1986) 'Late Holocene palaeoecology at Ketilstadir in Myrdalur, south Iceland', *Jokull* 36: 41–55.

Buntley, G.J. and Westin, F.C. (1965) 'A comparative study of developmental color in a chestnut–chernozem–brunizem soil climosequence', *Proceedings of the Soil Science Society of America* 29: 579–82.

Burns, I.G. (1974) 'A model for predicting the redistribution of salts applied to fallow soils after excess rainfall or evaporation', *Journal of Soil Science* 25: 165–78.

Burns, S.F. and Tonkin, P.J. (1982) 'Soil-geomorphic models and the spatial distribution and development of alpine soils', in *Space and Time in Geomorphology*, London: Allen & Unwin, pp.2–43.

Butler, B.E. (1955) 'A system for the description of soil structure and consistence in the field', *Journal of the Australian Institute of Agricultural Science* 21: 239–49.

Buurman, P. (1985) 'Carbon–sesquioxide ratios in organic complexes and the transition albic–spodic horizon', *Journal of Soil Science* 36: 255–60.

Cameron, D.R., Kowalenko, C.G. and Ivarson, K.C. (1978) 'Nitrogen and chloride leaching in a sandy field plot', *Soil Science* 126: 174–80.

Canada Soil Survey Committee (1978) The Canadian System of Soil Classification, Research Branch, Canadian Department of Agriculture, Publication 1646.

Carson, R. (1963) *Silent Spring*, London: Hamish Hamilton.

Carter, C.A. and Chorley, R.J. (1961) 'Early slope development in an expanding stream system', *Geological Magazine* 98: 117–30.

Catt, J.A. (1986) *Soils and Quaternary Geology: A Handbook for Field Scientists*, Oxford: Clarendon Press.

Chalmers, A.G. (1985) 'Review of information on lime loss and changes in soil pH gained from ADAS experiments', *Soil Use and Management* 1: 17–19.

Christopherson, R.W. (1992) *Geosystems: An Introduction to Physical Geography*, New York: Macmillan.

Churchman, G.J. and Tate, K.R. (1987) 'Stability of aggregates of different size grades in allophane soils from volcanic ash in New Zealand', *Journal of Soil Science* 38: 19–27.

Ciolkosz, E.J., Cronce, R.C., Cunningham, R.L. and Petersen, G.W. (1986) 'Geology and soils of Nittany Valley', *Pennsylvania State University Agronomy Series* 88, 52pp.

Ciolkosz, E.J., Carter, B.J., Hoover, M.T., Cronce, R.C., Waltman, W.J. and Dobos, R.R. (1990) 'Genesis of soils and landscapes in the ridge and valley province of central Pennsylvania', *Geomorphology* 3: 245–61.

Collins, V.G., D'Sylva, B.T. and Latter, P.M. (1976) 'Microbial populations in peat', in Heal, O.W. and Perkins, D.F. (eds) *Production Ecology of British Moors and Montane Grasslands*, Studies in Ecology 27: 94–112.

Conacher, A.J. and Dalrymple, J.B. (1977) 'The nine unit landsurface model: an approach to pedogeomorphic research', *Geoderma* 18: 1–154.

Conacher, A.J. and Dalrymple, J.B. (1978) 'Identification, measurement and interpretation of some pedogeomorphic processes', *Zeitschrift für Geomorphologie Supplementband* 29: 1–9.

Conaway, A.W. and Strickling, E. (1962) 'A comparison of selected methods for expressing soil aggregate stability', *Proceedings of the Soil Science Society of America* 24: 426–30.

Cooke, R.U. and Smalley, I.J. (1968) 'Salt weathering in deserts', *Nature* 220: 1226–7.

Cowan, J.A., Humphreys, G.S., Mitchell, P.B. and Murphy, C.L. (1985) 'An assessment of pedoturbation by two species of mound building ants, *Camponotus intrepidus* (Kirby) and *Iridomymex purpureus* (F. Smith)', *Australian Journal of Soil Research* 22: 98–108.

CPCS (Commission de Pédologie et de Cartographie des Sols) (1967) *Classification des sols*, Grignon: Ecole Nationale Supérieure Agronomique.

Creemans, D.L., Darmody, R.G. and Norton, L.D. (1992) 'Etch-pit size and shape distribution on orthoclase and pyriboles in a loess catena', *Geochimica et Cosmochimica Acta* 56: 3423–34.

Crompton, E. (1966) *The Soils of the Preston District of Lancashire*, Memoir of the Soil Survey of Great Britain, Harpenden.

Crowther, E.M. (1953) 'The sceptical soil chemist', *Journal of Soil Science* 4: 107–22.

Culbard, E.B., Thornton, I., Watt, J., Wheatley, M., Moorcroft, S. and Thompson, M. (1988) 'Metal contamination of British suburban dusts and soils', *Journal of Environmental Quality* 17: 226–34.

Curtis, C.D. (1976) 'Stability of minerals in surface weathering reactions', *Earth Surface Processes* 1: 63–70.

Dalrymple, J.B., Blong, R.J. and Conacher, A.J. (1968) 'A hypothetical nine-unit landsurface model', *Zeitschrift für Geomorphologie* 12: 60–76.

Dalziel, T.R.K., Proctor, M.V. and Dickson, A. (1988) 'Hydrochemical budget calculations for parts of the Loch Fleet catchment before and after watershed liming', *Water, Air and Soil Pollution* 41: 417–34.

Dan, J. and Yaalon, D.H. (1968) 'Pedomorphic forms and pedomorphic surfaces', *Transactions of the Ninth International Congress of Soil Science*, Adelaide 4: 577–84.

Darmody, R.G., Fanning, D.S., Drummond Jr, W.J. and Foss, J.E. (1977) 'Determination of total sulfur in tidal marsh soils by X-ray spectroscopy', *Journal of the Soil Science Society of America* 41: 761–5.

Dawson, H.J., Ugolini, F.C., Hrutfiord, B.F. and Zachara, J. (1978) 'Role of soluble organics in the soil processes of a podzol, central Cascades, Washington', *Soil Science* 126: 290–6.

de Bakker, H. and Schelling, J. (1966) *Systeem van bodem classificatie voor Nederland: de hogere niveaus*, Wageningen: Pudoc.

de Cornink, F. (1980) 'Major mechanisms in the formation of spodic horizons', *Geoderma* 24: 101–28.

Dixit, S.P., Gombeer, R. and D'Hoore, J. (1975) 'The electrophoretic mobility of natural clays and their potential mobility within the pedon', *Geoderma* 13: 325–30.

DoE/NWC (1981) *Report of the Sub-committee on the Disposal of Sewage to Land*, Standing Technical Committee Report No. 20, Department of the Environment, London.

Dokuchaev, V.V. (1898) *The Problem of the Re-evaluation of the Land in European and Asiatic Russia* (in Russian), Moscow.

Dowdell, R.J., Smith, K.A., Crees, R. and Restall, S.W.F. (1972) 'Field studies of ethylene in the soil atmosphere – equipment and preliminary results', *Soil Biology and Biochemistry* 4: 325–31.

Du Bois, G.B. and Jeffrey, P.G. (1955) The compositions and origins of the laterites of the Entebbe Peninsula, Uganda Protectorate, *Colonial Geology and Mineral Resources* 5: 387–408.

Duchaufour, P. (1982) *Pedology: Pedogenesis and Classification*, translated from the French by T.R. Paton, London: Allen & Unwin.

Dunn, J.R. and Hudec, P.P. (1966) 'Clay, water and rock soundness', *Ohio Journal of Science* 66: 153–68.

Dunn, J.R. and Hudec, P.P. (1972) 'Frost absorption in argillaceous rocks', in *Frost Action in Soils*, Highway Research Record 393, Highway Research Board, pp.65–78.

Dunne, T. (1978) 'Field studies of hillslope processes', in M.J. Kirkby (ed.) *Hillslope Hydrology*, Chichester: J. Wiley & Sons.

Ellis, S. and Mellor, A. (1995) *Soils and Environment*, London: Routledge.

Ellison, W.D. (1945) 'Some effects of raindrops and surface-flow on soil erosion and infiltration', *Transactions of the American Geophysical Union* 26: 415–29.

England, C.B. and Holtan, H.N. (1969) 'Geomorphic grouping of soils in watershed engineering', *Journal of Hydrology* 7: 217–25.

Evans, R. (1988) *Water Erosion in England and Wales. 1982–1984*, Report for Soil Survey and Land Resource Centre, Silsoe.

Evans, R. and Cook, S. (1986) 'Soil erosion in Britain', *SEESOIL* 3: 28–59.

Fahey, B.D. (1983) 'Frost action and hydration as rock weathering mechanisms on schist: a laboratory study', *Earth Surface Processes and Landforms* 8: 535–45.

Fanning, D.S. and Fanning, M.C.B. (1989) *Soil: Morphology, Genesis and Classification*, New York: J. Wiley & Sons.

FAO/UNESCO (1974) *Soil Map of the World: Volume 1. legend*, Paris: UNESCO.

FAO/UNESCO (1989) *Soil Map of the World: revised legend*, Wageningen: International Soil Reference and Information Centre.

Farmer, V.C. (1982) 'Significance of the presence of allophane and imogolite in podzol B_s horizons for podzolisation mechanisms: a review', *Soil Science and Plant Nutrition* 28(4): 571–8.

Farmer, V.C. and Fraser, A.R. (1982) 'Chemical and colloidal stability of sols in the Al_2O_3–Fe_2O_3–SiO_2–H_2O system: their role in podzolisation', *Journal of Soil Science* 33: 737–42.

Farmer, V.C., Russell, J.D. and Berrow, M.L. (1980) 'Imogolite and proto-imogolite allophane in spodic horizons: evidence for a mobile aluminium silicate complex in podzol formation', *Journal of Soil Science* 31: 673–84.

Farres, P. (1978) 'The role of time and aggregate size in the crusting process', *Earth Surface Processes and Landforms* 3: 243–54.

Fenneman, N.M. (1916) 'Physiographic divisions of the United States', *Annals of the Association of American Geographers* 6: 19–98.

Fisher, G. and Yan, O. (1984) 'Iron mobilisation by heathland plant extracts', *Geoderma* 32: 339–45.

FitzPatrick, E.A. (1971) *Pedology: A Systematic Approach to Soil Science*, Edinburgh: Oliver & Boyd.

FitzPatrick, E.A. (1983) *Soils: Their Formation. Classification and Distribution*, 2nd edn, London: Longman.

Flury, M., Flühler, H., Jury, W.A. and Leuenberger, J. (1994) 'Susceptibility of soils to preferential flow of water: a field study', *Water Resources Research* 30: 1945–54.

Fogel, R. and Cromack, K. (1977) 'Effect of habitat and substrate quality on Douglas fir litter decomposition in western Oregon', *Canadian Journal of Botany* 55: 1632–40.

Foster, W.D.A. (1950) 'Comparison of 9 indices of rainfall intensity', *Transactions of the American Geophysical Union* 31: 894–900.

Foth, H.D. and Schafer, J.W. (1980) *Soil Geography and Land Use*, New York: J. Wiley & Sons.

Fowler, D., Cape, J.N. and Leith I.D. (1985) 'Acid

inputs from the atmosphere in the United Kingdom', *Soil Use and Management* 1: 3–5.

Frenkel, H., Gertsl, Z. and Alperovitch, N. (1989) 'Exchange-induced dissolution of gypsum and the reclamation of sodic soils', *Journal of Soil Science* 40: 599–611.

Friedman, H. (1985) 'The science of global change – an overview', in Malone, T.F. and Roederer, J.G. (eds), *Global Change*, Cambridge: Cambridge University Press.

Frostick, L.E. and Reid, I. (1982) 'Alluvial processes, mass wasting and slope evolution in arid environments', *Zeitschrift für Geomorphologie Supplementband* 44: 53–67.

Fullen, M.A. (1985a) 'Wind erosion of arable soils in east Shropshire (England) during spring 1983', *Catena* 12: 111–20.

Fullen, M.A. (1985b) 'Compaction, hydrological processes and soil erosion on loamy sands in east Shropshire, England', *Soil and Tillage Research* 6: 17–29.

Furley, P.A. (1968) 'Soil formation and slope development: 2 The relationship between soil formation and gradient in the Oxford area', *Zeitschrift für Geomorphologie* NF 12: 25–42.

Furley, P.A. (1971) 'Relationships between slope form and soil properties developed over chalk parent materials', in Brunsden, D. (ed.) *Slopes, Form and Process*, Institute of British Geographers Special Publication 3, pp.141–64.

Gardiner, M.J. and Radford, T. (1980) *Soil Associations of Ireland and their Land Use Potential*, Soil Survey Bulletin No. 36, National Soil Survey of Ireland, An Foras Talúntais, Dublin.

Garron, S. and Hadas, A. (1973) 'Measurement of the water status of soil and soil characteristics relevant to irrigation', in Yaron, B., Danfors, E. and Vaadia, Y (eds) *Arid Zone Irrigation*, London: Chapman & Hall, pp.215–40.

Gerrard, A.J. (1982) 'Slope form and regolith characteristics in the basin of the River Cowsic, Central Dartmoor, Devon', PhD thesis, University of London.

Gerrard, A.J. (1988) *Rocks and Landforms*, London: Allen & Unwin.

Gerrard, A.J. (1989) 'The nature of slope materials on the Dartmoor granite', *Zeitschrift für Geomorphologie* NF 33: 179–88.

Gerrard, A.J. (1990) 'Soil variations on hillslopes in humid temperate climates', *Geomorphology* 3: 225–44.

Gerrard, A.J. (1992a) *Soil Geomorphology*, London: Chapman & Hall.

Gerrard, A.J. (1992b) 'The nature and geomorphological relationships of earth hummocks (thufur) in Iceland', *Zeitschrift für Geomorphologie Supplementband* 86: 169–78.

Gerrard, A.J. (1994) 'Weathering of granitic rocks: environment and clay mineral formation', in Robinson, D.A. and Williams, R.B.G. (eds) *Rock Weathering and Landform Evolution*, Chichester: J. Wiley & Sons, pp.3–20.

Gersper, P.L. and Holowaychuk, N. (1970a) 'Effects of stemflow water on a Miami soil under a beech tree: I. Morphological and physical properties', *Proceedings of the Soil Science Society of America* 34: 779–86.

Gersper, P.L. and Holowaychuk, N. (1970b) 'Effects of stemflow water on a Miami soil under a beech tree: II. Chemical properties', *Proceedings of the Soil Science Society of America* 34: 786–94.

Gersper, P.L. and Holowaychuk, N. (1971) 'Some effects of stem flow from forest canopy trees on chemical properties of soils', *Ecology* 52: 691–702.

Ghilarov, M.S. (1970) 'Soil biocoenoses', in Phillipson, J. (ed.) *Methods of Study in Soil Ecology*, UNESCO, pp.67–77.

Gibbs, H.S. (1980) *New Zealand Soil: An Introduction*, Wellington, New Zealand: Oxford University Press.

Gile, L.H. and Grossman, R.B. (1968) 'Morphology of the argillic horizon in desert soils of southern New Mexico', *Soil Science* 106: 6–15.

Gile, L.H. and Grossman, R.B. (1979) *The Desert Project Soil Monograph*, US Department of Agriculture, Soil Conservation Service, Washington, DC.

Gile, L.H., Hawley, J.W. and Grossman, R.B. (1981) 'Soils and geomorphology in the basin and range area of southern New Mexico – guidebook to the Desert Project', *New Mexico Bureau of Mines and Minerals Resources*, memoir 39.

Gile, L.H. and Hawley, J.W. (1968) 'Age and comparative development of desert soils at the Gardner Spring radiocarbon site, New Mexico', *Proceedings of the Soil Science Society of America* 32: 709–16.

Gilkes, R.J., Scholz, G. and Dimmock, G.M. (1973) 'Lateritic deep weathering of granite', *Journal of Soil Science* 24: 523–36.

Gilman, K. and Newson, M.D. (1980) *Soil Pipes and Pipeflow*, British Geomorphological Research Group Research Monograph 1, Norwich: Geo Abstracts.

GLASOD (Global soils degradation data base) (in Middleton and Thomas, 1997).

Goldich, S.S. (1938) 'A study of rock weathering', *Journal of Geology* 46: 17–58.

Goodman, R.E. (1980) *Introduction to Rock Mechanics*, New York: J. Wiley & Sons.

Goudie, A.S. (1973) *Duricrusts in Tropical and Subtropical Landscapes*, Oxford: Clarendon Press.

Goudie, A.S. (1974) 'Further experimental investigation of rock weathering by salt and other mechanical processes', *Zeitschrift für Geomorphologie Supplementband* 21: 1–12.

Gray, T.R.G. and Williams, S.T. (1971) *Soil Microorganisms*, Edinburgh: Oliver & Boyd.

Green, P. (1974) 'Recognition of sedimentary characteristics in soils by size–shape analysis', *Geoderma* 11: 181–93.

Grieve, I.C. (1985) 'Annual losses of iron from moorland soils and their relation to free iron contents', *Journal of Soil Science* 36: 307–12.

Griggs, D.T. (1936) 'The factor of fatigue in rock weathering', *Journal of Geology* 44: 781–96.

Gustafson, D.I. (1989) 'Groundwater ubiquity score: a simple method for assessing pesticide leachability', *Environmental Toxicology and Chemistry* 8: 339–57.

Gwynne, C.S. and Simonson, R.W. (1942) 'Influence of low recessional moraines on soil type pattern of the Mankato drift plain in Iowa', *Soil Science* 53: 461–6.

Hack, J.T. and Goodlett, J.G. (1960) *Geomorphology and Forest Ecology of a Mountain Region in the Central Appalachians*, US Geological Survey Professional Paper 347.

Haigh, R.A. and White, R.E. (1986) 'Nitrate leaching from a small, underdrained, grassland, clay catchment', *Soil Use and Management* 2: 65–70.

Hallbacken, L. and Tamm, C.O. (1986) 'Changes in soil acidity from 1927 to 1982–1984 in a forest area of southwest Sweden', *Scandinavian Journal of Forest Research* 1: 219–32.

Hammond, R.F. (1981) *The Peatlands of Ireland*, 2nd edn, Soil Survey Bulletin No. 35, National Soil Survey of Ireland, An Foras Talúntais, Dublin.

Harden, J.W. (1982) 'A quantitative index of soil development from field descriptions: examples from a chronosequence in central California', *Geoderma* 28: 1–28.

Harden, J.W. and Taylor, E.M. (1983) 'A quantitative comparison of soil development in four climatic regimes', *Quaternary Research* 20: 342–59.

Harris, C. (1983) 'Vesicles in thin sections of periglacial soils from north and south Norway, *Proceedings Fourth International Conference on Permafrost*, Fairbanks, Alaska, 445–9.

Haynes, R.J. and Swift, R.S. (1990) 'Stability of soil aggregates in relation to organic constituents and soil water content', *Journal of Soil Science* 41: 73–83.

Heal, O.W. and French, D.D. (1974) 'Decomposition of organic matter in tundra', in Holding, A.J., Heal, O.W., MacLean, S.F. and Flanagan, P.W. (eds) *Soil Organisms and Decomposition in Tundra*, IBP Tundra Biome, pp.279–309.

Heal, O.W., Latter, P.M. and Howson, G. (1978) 'A study of the rates of decomposition of organic matter', in Heal, O.W. and Perkins, D.F. (eds) *Production Ecology of British Moors and Montane Grasslands*, Ecology Studies 27, New York: Springer Verlag, pp.136–59.

Heede, B.H. (1971) *Characteristics and Processes of Soil Piping in Gullies*, US Department of Agriculture and Forest Services Research Paper, RM-68.

Helgeson, H.C. (1971) 'Kinetics of mass transfer among silicates and aqueous solutions', *Geochimica et Cosmochimica Acta* 35: 4121–68.

Herath, J.W. and Grimshaw, R.W. (1971) 'A general evaluation of the frequency distribution of clay and associated minerals in the alluvial soils of Ceylon', *Geoderma* 5: 119–30.

Herbillon, A.J. and Nahon, D.D. (1988) 'Laterites and lateritization processes', in Stuaki, J.W., Goodman, B.A. and Schwertmann, V. (eds) *Iron in Soils and Clay Minerals*, Dordrecht: Reidel.

Hewitt, A.E. (1992) *New Zealand Soil Classification*, DSIR Land Resources Scientific Report No. 19.

Hewlett, J.D. and Nutter, W.L. (1970) 'The varying source area of streamflow from upland basins', paper presented at symposium on interdisciplinary aspects of watershed management, Montana State University, Bozeman, New York, American Society of Civil Engineers.

Hillel, D. (1992) *Out of the Earth: Civilization and the Life of the Soil*, London: Aurum Press.

Hogan, D.V. (1982) 'Upland soil survey in southwest England', Welsh Soils Discussion Group Report 20: 135–61.

Hole, F.D. (1981) 'Effects of animals on soils', *Geoderma* 25: 75–112.

Holmes, D.A. and Western, S. (1969) 'Soil-texture patterns in the alluvium of the lower Indus Plains', *Journal of Soil Science* 20: 23–37.

Hornung, M. (1985) 'Acidification of soils by trees and forests', *Soil Use and Management* 1: 24–8.

Hornung, M., Reynolds, B., Stevens, P.A. and Neal, C. (1987) 'Increased acidity and aluminium concentrations in streams following afforestation', in *Proceedings of the International Symposium on Acidification and Water Pathways*, Norwegian National Committee for Hydrology, Bolkesjo, Norway, pp.259–268.

Huggett, R.J. (1995) *Geoecology*, London: Routledge.

Humphreys, G.S. (1981) 'The rate of ant mounding and earthworm casting near Sydney, New South Wales', *Search* 12: 129–31.

Humphreys, G.S. (1985) 'Bioturbation, rainwash and texture contrast soils', unpublished PhD thesis, School of Earth Sciences, Macquarie University, Australia.

Humphreys, G.S. (1994) 'Bioturbation, biofabrics and the biomantle: an example from the Sydney basin', in Ringrose-Voase, A.J. and Humphreys, G.S. (eds) *Soil Micromorphology: Studies in Management and Genesis*, Amsterdam: Elsevier, pp.421–36.

Humphreys, G.S. and Mitchell, P.B. (1983) 'A preliminary assessment of the role of bioturbation and rainwash on sandstone hillslopes in the Sydney basin', in Young, R.W. and Nanson, G.C. (eds) *Aspects of Australian Sandstone Landscape*, Wollongong: University of Wollongong, pp.65–80.

Hurst, V.J. (1977) 'Visual estimation of iron in saprolite', *Bulletin of the Geological Society of America* 88: 174–6.

Ilaiwi, M., Abdelgawad, G. and Jabour, E. (1992) 'Syria: human-induced soil degradation', in *UNEP World Atlas of Desertification*, London: Edward Arnold, pp.42–5.

Imeson, A.C. (1995) 'The physical, chemical and

biological degradation of the soil', in Fantechi, R., Peter, D., Balabanis, P. and Rubio, J.L. (eds) *Desertification in a European Context: Physical and Socio-economic Aspects*, Proceedings of the European School of Climatology and Natural Hazards, Alicante, Spain, 6–13 October, 1993, Brussels: European Community, pp.153–68.

Imeson, A.C. and Jungerius, P.D. (1976) 'Aggregate stability and colluviation in the Luxembourg Ardennes: An experimental and micromorpho-logical study', *Earth Surface Processes and Landforms* 1: 259–71.

Iraq, government of (1980) Desertification in the Greater Mussayeb project, Iraq, in *UNESCO/UNEP/UNDP Case Studies on Desertification*, Natural Resources Research Series XVIII, Paris: UNESCO, pp.176–213.

Isbell, R.F. (1992) 'A brief history of national soil classification in Australia since the 1920s', *Australian Journal of Soil Research* 30: 825–42.

Isbell, R.F. (1995) 'The use of sodicity in Australian soil classification systems, in Naidu, R., Sumner, M.E. and Rengasamy, P. (eds) *Australian Sodic Soils: Distribution, Properties and Management*, Melbourne: CSIRO Publishing, pp.41–6.

Isbell, R.F., McDonald, W.S. and Ashton, L.J. (1997) *Concepts and Rationale of the Australian Soil Classification*, Canberra: ACLEP, CSIRO Land and Water.

Jahn, A. (1963) 'Importance of soil erosion for the evolution of slopes in Poland', *Nach. Akad. Wiss. Gottingen Math-Phys Kl.* 15, 229–37.

James, P.A. (1970) 'The soils of the Rankin Inlet, Keewatin, NWT, Canada', *Arctic and Alpine Research* 2: 293–302.

Jenkinson, D.S. (1990) 'The turnover of organic carbon and nitrogen in soil', *Philosophical Transactions of the Royal Society Series B* 329: 361–8.

Jenny, H. (1941) *Factors of Soil Formation. A System of Quantitative Pedology*, New York: McGraw-Hill.

Jenny, H. (1961) 'Derivation of state factor equations of soils and ecosystems', *Proceedings of the Soil Science Society of America* 25: 385–8.

Johnson, D.L. and Watson-Stegner, D. (1987) 'Evolution model of pedogenesis', *Soil Science* 143: 349–66.

Jones, R.L. and Beavers, H.A. (1964) 'Variation of opal phytolith content among some great soil groups of Illinois', *Proceedings of the Soil Science Society of America* 28: 711–12.

Jury, W.A., Gardner, W.R., Saffiena, P.G. and Tanner, C.B. (1976) 'Model for predicting simul-taneous movement of nitrate and water through a loamy sand', *Soil Science* 122: 36–43.

Jury, W.A., Spencer, W.F. and Farmer, W.J. (1983) 'Behaviour assessment model for trace organics in soil. I: Model desorption', *Journal of Environment Quality* 12: 558–64.

Khoshoo, T.N. and Tejwani, K.G. (1993) 'Soil erosion and conservation in India (status and policies)', in Pimentel, D. (ed.) *World Soil Erosion and Conservation*, Cambridge: Cambridge University Press, pp.109–45.

Kirkby, M.J. (1969) 'Erosion by water on hillslopes', in Chorley, R.J. (ed.) *Water, Earth and Man*, London: Methuen.

Kirkby, M.J. and Chorley, R.J. (1967) 'Through-flow, overland flow and erosion', *Bulletin of the International Association of Scientific Hydrologists* 12: 5–21.

Klemmenson, J.O. (1987) 'Influence of oak in pine forests of central Arizona on selected nutrients of forest floor and soil', *Journal of the Soil Science Society of America* 51: 1623–8.

Krinsley, D.H. and Doornkamp, J. (1973) *Atlas of Quartz Sand Surface Textures*, Cambridge: Cambridge University Press.

Krishnamoorthy, R.V. (1985) 'A comparative study of wormcast production by earthworm populations from grassland and woodland near Bangalore, India', *Revue d'Ecologie et de Biologie du Sol* 22: 209–19.

Lal, R. and Stewart, B.A. (1990) 'Need for action: research and development priorities', *Advances in Soil Science* 11: 331–6.

Langford-Smith, T. (ed.) (1978) *Silcrete in Australia*, Department of Geography, University of New England, Armidale, Australia.

Lavado, R.S. and Taboada, M.A. (1987) 'Soil salinization as an effect of grazing in a native grassland soil in the Flooding Pampa of Argentina', *Soil Use and Management* 3: 143–8.

Lavelle, P. (1978) *Les Vermes de terre de la savane de Lamto (Côte d'Ivoire) peuplements, populations et fonctions dans l'ecosysteme* (Publications du Laboratoire de Zoologie 12), Paris: Ecole Normale Supérieure.

Lee, K.E. (1985) *Earthworms: Their Ecology and Relationships with Soils and Land Use*, Sydney: Academic Press.

Lietzke, D.A. and McGuire, G.A. (1987) 'Characterization and classification of soils with spodic morphology in the southern Appalachians', *Journal of the Soil Science Society of America* 51: 165–70.

Linton, D.L. (1951) 'The delimitations of morphological regions', in Stamp, L.D. and Wooldridge, S.W. (eds) *London Essays in Geography*, London: Longman, pp.199–217.

Litaaor, M.I., Mancinelli, R. and Halfpenny, J.C. (1996) 'The influence of pocket gophers on the status of nutrients in alpine soils', *Geoderma* 70: 37–48.

Livens, F.R. and Loveland, P.J. (1988) 'The influence of soil properties on the environmental mobility of caesium in Cumbria', *Soil Use and Management* 4: 69–75.

Lockeretz, W. (1978) 'The lessons of the Dust Bowl', *American Scientist* 66: 560–9.

Lundstrom, U.S. (1993) 'The role of organic acids in the soil solution chemistry of a podzolised soil', *Journal of Soil Science* 44: 121–33.

Machette, M.N. (1985) 'Calcic soils of the southwestern United States', *Geological Society of America, Special Paper* 203: 1–21.

Macklin, M.G. (1986) 'Channel and floodplain metamorphosis in the River Nent, Cumberland', in Macklin, M.G. and Rose, J. (eds) *Quaternary River Landforms and Sediments in the North Pennines*, BGRG/QRA, Cambridge, pp.19–33.

Mackney, D., Hodgson, J.M., Hollis, J.M. and Staines, S.J. (1983) Legend for the 1:250,000 Soil Map of England and Wales, Soil Survey of England and Wales, Harpenden.

Madge, D.S. (1969) 'Field and laboratory studies on the activities of two species of tropical earthworms', *Pedobiologia* 9: 188–214.

Malcolm, R.L. and McCracken, R.J. (1968) 'Canopy drip: a source of mobile soil organic matter for mobilisation of iron and aluminium', *Proceedings of the Soil Science Society of America* 32: 834–8.

Mando, A., Stoosnijder, L. and Brussaars, L. (1996) 'Effects of termites on infiltration into crusted soil', *Geoderma* 74: 107–13.

Mann, A.W. and Ollier, C.D. (1985) 'Chemical diffusion and ferricrete formation', *Soils and Geomorphology, Catena Supplement* 6: 151–7.

McCall, P.J., Laskowski, D.A., Swann, R.L. and Dishburger, H.J. (1980) 'Measurement of sorption coefficients of organic chemicals and their use in environmental fate analysis', in *Test Protocols for Environmental Fate and Movement of Toxicants*, Proceedings of a Symposium of the Association of Official Analytical Chemists, 94th Annual Meeting, 1980, Washington, DC, pp.89–109.

McDonald, E.V. and Busacca, A.J. (1990) 'Interaction between aggrading geomorphic surfaces and the formation of a Late Pleistocene paleosol in the Palouse loess of eastern Washington state', *Geomorphology* 3: 449–70.

McDonald, R.C., Isbell, R.F., Speight, J.G., Walker, J. and Hopkins, M.S. (1990) *Australian Soil and Land Survey Field Handbook*, 2nd edn, Melbourne: Inkata Press.

McFadden, L.D. and Tinsley, J.C. (1985) 'The rate and depth of accumulation of pedogenic carbonate accumulation in soils: formation and testing of a compartment model', in Weide, D.W. (ed.) *Soils and Quaternary Geology of the Southwestern United States*, Geological Society of America, Special Paper 203, pp.23–42.

McFarlane, M.J. (1976) *Laterite and Landscape*, London: Academic Press.

McGinnies, W. and Laycock, W. (1988) 'The Great American Desert, USA: perceptions of pioneers, the Dust Bowl and the new sodbusters', in Whitehead, E. (ed.) *Arid Lands Today and Tomorrow*, Boulder, Colo.: pp.1247–54.

McGreevy, J.P. (1982) 'Frost and salt weathering: further experimental results', *Earth Surface Processes and Landforms* 7: 475–88.

McGreevy, J.P. (1985) 'Thermal properties as controls on rock surface temperature maxima, and possible implications for rock weathering', *Earth Surface Processes and Landforms* 10: 125–36.

McIntosh, P.D. (1980) 'Weathering products in vitrandept profiles under pine and manuka, New Zealand', *Geoderma* 24: 225–39.

McKeague, J.A., Wang, C., Coen, G.M., DeKimpe, C.R., Laverdier, M.R., Evans, L.J., Kloosterman, B. and Green, A.J. (1983) 'Testing chemical criteria for spodic horizons on podzolic soils in Canada', *Journal of the Soil Science Society of America* 47: 1052–4.

McRae, S.G. and Burnham, C.P. (1976) 'Soil classification', *Classification Society Bulletin* 3: 56–64.

Meilhac, A. and Tardy, Y. (1970) 'Génése et évolution des sericites, vermiculites et montmorillonites au cours de l'alteration des plagioclases un pays témperé', *Bulletin Servis Carte Geologique, Alsace Lorraine* 23: 145–61.

Mellanby, K. (1967) *Pesticides and Pollution*, Fontana New Naturalist Series, London: Collins.

Mellor, A. (1985) 'Soil chronosequences on neoglacial moraine ridges, Jostedalsbreen and Jotunheimen, southern Norway: a quantitative pedogenic approach', in Richards, K.S., Arnett, R.R. and Ellis, S. (eds) *Geomorphology and Soils*, London: Allen & Unwin, pp. 289–308.

Melville, M.D. and Atkinson, G. (1985) 'Soil colour: its measurement and its designation in models of uniform colour space', *Journal of Soil Science* 36: 495–512.

Meyer, L.D. and Kramer, L.A. (1969) 'Erosion equations predict land slope development', *Agricultural Engineering* 50: 5222–3.

Middleton, H.E. (1930) *Properties of Soils which Influence Soil Erosion*, US Department of Agriculture Technical Bulletin No. 178, 1–16.

Middleton, N. and Thomas, D. (1997) *World Atlas of Desertification*, 2nd edn, London: Edward Arnold.

Miller, H.G. (1985) 'The possible role of forests in streamwater acidification', *Soil Use and Management* 1: 28–9.

Miller, J.P. (1961) 'Solutes in small streams draining single rock types, Sangre de Cristo range, New Mexico', US Geological Survey Water Supply Paper 1535F.

Milne, G. (1935a) 'Some suggested units of classification and mapping particularly for East Africa', *Soils Research* 4: No. 3.

Milne, G. (1935b) 'Composite units for the mapping of complex soil associations', *Transactions of the Third International Congress of Soil Science* 1: 345–7.

Milne, R. (1987) 'Acid soils are harbouring Chernobyl's caesium', *New Scientist* 115: 28.

Minderman, G. (1968) 'Addition, decomposition, and accumulation of organic matter in forests', *Journal of Ecology* 56: 355–62.

Mitchell, P.B. (1985) 'Some aspects of the role of bioturbation in soil formation in south-eastern Austalia', PhD thesis, School of Earth Sciences, Macquarie University, Australia.

MOA (1985) *Indian Agriculture in Brief*, 20th edn, Ministry of Agriculture, New Delhi, India.

Molope, M.B., Grieve, I.C. and Page, E.R. (1987) 'Contributions by fungi and bacteria to aggregate stability of cultivated soils', *Journal of Soil Science* 38: 71–7.

Moore, A.W., Isbell, R.F. and Northcote, K.H. (1983) 'Classification of Australian soils, in CSIRO, *Soils: An Australian Viewpoint*, London: Academic Press, pp.253–6.

Morgan, R.P.C. (1977) *Soil Erosion in the United Kingdom: Field Studies in the Silsoe Area, 1973–75*, Occasional Paper No. 4, National College of Agricultural Engineering, Silsoe.

Morgan, R.P.C. (1995) *Soil Erosion and its Conservation*, 2nd edn, Harlow: Longman.

Morin, J., Benyamini, Y. and Michaeli, A. (1981) 'The effect of raindrop impact on the dynamics of soil surface crusting and water movement in the profile', *Journal of Hydrology* 52: 321–6.

Moss, R.P. (1968) 'Soils, slopes and surfaces in tropical Africa', in *The Soil Resources of Tropical Africa*, Cambridge: Cambridge University Press, pp.29–60.

Mückenhausen, E. (1985) *Die Bodenkunde und ihre geologischen, geomorphologischen, mineralogischen une petrologischen Grundlagen* 3, ergänzte Auflage, Frankfurt am Main: DLG Verlag.

Mullins, C.E., Young, I.M., Bengough, A.G. and Ley, G.J. (1987) 'Hardsetting soils', *Soil Use and Management* 3: 79–83.

Musgrave, G.W. (1947) 'Quantitative evaluation of factors in water erosion – a first approximation', *Journal of Soil and Water Conservation* 2: 133–8.

Nikiforoff, C.C. (1949) 'Weathering and soil formation', *Soil Science* 67: 219–30.

Nortcliff, S., Quisenberry, V.L., Nelson, P. and Phillips, R.E. (1994) 'The analysis of soil macropores and the flow of solutes', in Humphreys, G. and Ringrose-Vaase, A.J. (eds) *Soil Micromorphology*, Proceedings of the Ninth International Working Meeting on Soil Micromorphology, Townsville, Queensland, 1992, pp.601–12.

Northcote, R.H. (1979) *A Factual Key for the Recognition of Australian Soils*, 4th edn, Adelaide: Rellim Technical Publication.

Norton, E.A. and Smith, R.S. (1930) 'Influence of topography on soil profile character', *Journal of the American Society of Agronomy* 22: 251–62.

Nyberg, P. and Thornelof, E. (1988) 'Operational liming of surface waters in Sweden', *Water, Air and Soil Pullution* 41: 3–16.

O'Hara, S.L., Street-Perrot, F.A. and Burt, T.P. (1993) 'Accelerated soil erosion around a Mexican highland lake caused by prehistoric agriculture', *Nature* 362: 48–51.

Oldeman, L.R. (1988) *Guidelines for General Assessment of the Status of Human-induced Soil Degradation*, Wageningen: ISRIC.

Oldeman, L.R., Hakkeling, R.T.A. and Sombroek, W.G. (1990) *World Map of the Status of Human-induced Soil Degradation: An Explanatory Note*, Wageningen: ISRIC and Nairobi: UNEP.

Oliver, M.A. and Gerrard, A.J. (1996) 'Soil erosion in the West Midlands', in Gerrard, A.J. and Slater, T.R. (eds) *Managing a Conurbation: Birmingham and its Region*, Studley: Brewin Books, pp.46–58.

Oliver, M.A. and Webster, R. (1987) 'The elucidation of soil patterns in the Wyre Forrest of the West Midlands, England. 1. Multivariate distribution', *Journal of Soil Science* 38: 279–91.

Ollier, C.D. (1959) 'A two-cycle theory of tropical pedology', *Journal of Soil Science* 10: 137–48.

Ollier, C.D. (1976) 'Catenas in different climates', in Derbyshire, E. (ed.) *Geomorphology and Climate*, Chichester: J. Wiley & Sons.

Ollier, C.D. (1984) *Weathering*, 2nd edn, London: Longman.

Ollier, C.D. and Pain, C. (1996) *Regolith, Soils and Landforms*, Chichester: J. Wiley & Sons.

Ormerod, S.J., Donald, A.P. and Brown, S.J. (1989) 'The influence of plantation forestry on the pH and aluminium concentrations of upland Welsh streams: a re-examination', *Environmental Pollution* 62: 47–62.

Parker, G.G. (1964) *Piping. A Geomorphic Agent in Landform Development of the Drylands*, International Association of Scientific Hydrology Publication 65.

Parton, W.J., Schimel, D.S., Cole, C.V. and Ojima, D.S. (1987) 'Analysis of factors controlling soil organic matter levels in Great Plains grassland', *Journal of the Soil Science Society of America* 51: 1173–9.

Parton, W.J., Stewart, J.W.B. and Cole, C.V. (1988) 'Dynamics of C, N, P and S in grassland soils: a model', *Biogeochemistry* 5: 109–31.

Paton, T.R., Humphreys, G.S. and Mitchell, P.B. (1995) *Soils: A New Global View*, London: UCL Press.

Patrick, W.H., Jr (1978) 'Critique of "Measurement and Prediction of Anaerobics in Soils"', in Neilson, D.R. and MacDonald, J.G. (eds) *Nitrogen in the Environment, Vol. 1. Nitrogen Behaviour in Field Soil*, London: Academic Press, pp.449–57.

Pennock, D.J. and de Jong, E. (1990) 'Rates of soil distribution associated with soil zones and slope classes in southern Saskatchewan', *Canadian Journal of Soil Science* 70: 325–34.

Peterson, F.F. (1980) 'Holocene desert soil formation under sodium salt influence in a playa-margin environment', *Quaternary Research* 13: 172–86.

Piccolo, A. and Mbagwu, J.S.C. (1990) Effects of different organic waste amendments on soil microaggregate stability and molecular sizes of humic substances', *Plant and Soil* 123: 27–37.

Polunin, N. and Grinevald, J. (1988) 'Vernadsky and biospheral ecology', *Environmental Conservation* 15: 117–22.

Polynov, B. (1937) *The Cycle of Weathering*, London: Murby.

Pons, L.J. and Zonneveld, I.S. (1965) *Soil Ripening and Soil Classification: Initial Soil Formation of Alluvial Deposits with a Classification of the Resulting Soils*, Wageningen International Institute for Land Reclamation and Improvement, Wageningen.

Pullen, R.A. (1979) 'Termite hills in Africa; their characteristics and evolution', *Catena* 6: 267–91.

Pye, (1987) *Aeolian Dust and Dust Deposits*, London: Academic Press.

Reed, A.H. (1979) 'Accelerated erosion of arable soils in the U.K. by rainfall and run-off', *Outlook on Agriculture* 10: 41–8.

Reed, A.H. (1983) 'The erosion risk of compaction', *Soil and Water* 11: 29–33.

Reed, G., Kemp, R.A. and Rose, J. (1996) 'Development of a feldspar weathering index and its application to a buried soil chronosequence in southeastern England', *Geoderma* 74: 267–280.

Reynolds, R.C., Jr (1971) 'Clay mineral formation in an alpine environment', *Clays and Clay Minerals* 19: 361–74.

Rieger, S. (1974) 'Arctic soils', in Ives, J.D. and Barry, R.G. (eds) *Arctic and Alpine Environments*, London: Methuen, pp.749–69.

Ross, S. (1989) *Soil Processes: A Systematic Approach*, London: Routledge.

Rowell, D.L. (1994) *Soil Science: Methods and Applications*, Harlow: Longman.

Rozov, N.N. and Ivanova, E.N. (1967) 'Classification of the soils of the USSR', *Soviet Soil Science* 2: 147–56.

Rubin, J. (1966) 'Theory of rainfall uptake by soils initially drier than their field capacity and its application', *Water Resources Research* 2: 739–94.

Rubio, J.L. (1995) 'Desertification: evolution of a concept', in Fantechi, R., Peter, D., Balabanis, P. and Rubio, J.L. (eds) *Desertification in a European Context: Physical and Socio-economic Aspects*, Proceedings of the European School of Climatology and Natural

Hazards Course, Alicante, Spain, 6–13 October 1993, Brussels: European Commission, pp.5–13.

Ruhe, R.V. (1956) 'Geomorphic surfaces and the nature of soils', *Soil Science* 82: 441–55.

Ruhe, R.V. (1969) *Quaternary Landscapes in Iowa*, Ames, Iowa: Iowa State University Press.

Runge, E.C.A. (1973) 'Soil development sequences and energy models', *Soil Science* 115: 183–93.

Russell, E.W. (1973) *Soil Conditions and Plant Growth*, 10th edn, London: Longman.

Rutter, N.W., Foscolos, A.E. and Hughes, O.L. (1978) 'Climatic trends during the Quaternary in central Yukon based upon pedological and geomorphological evidence', in Mahaney, W.C. (ed.) *Quaternary Soils*, Norwich: Geo Books.

Ruxton, B.P. and Berry, L. (1957) 'Weathering of granite and associated features in Hong Kong', *Bulletin of the Geological Society of America* 68: 1263–92.

Satchell, J.E. (1958) 'Earthworm biology and soil fertility', *Soils and Fertilizers* 21: 209–19.

Scharpenseel, H.W. and Kerpen, W. (1967) 'Studies on tagged clay migration due to water movement', in *Proceedings International Atomic Energy Agency Symposium*, Istanbul, Vienna: IAEA, pp.287–90.

Schnitzer, M. and Desjardins, J.G. (1969) 'Chemical characteristics of a natural soil leachate from a humic podzol', *Canadian Journal of Soil Science* 49: 151–8.

Schnitzer, M. and Skinner, S.I.M. (1963) 'Organo-metallic interactions in soils. 1. Reactions between a number of metal ions and the organic matter of a podzol B_h horizon,' *Soil Science* 96: 86–93.

Schnitzer, M. and Skinner, S.I.M. (1965) 'Organo-metallic interactions in soils. 4. Carboxyl and hydroxyl groups in organic matter and metal retention', *Soil Science* 99: 278–84.

Scrivner, C.L., Bake, J.C. and Brees, D.R. (1973) 'Combined daily climatic data and dilute solute chemistry in studies of soil profile formation', *Soil Science* 115: 213–23.

Shaw, C.F. (1930) 'Potent factors in soil formation', *Ecology* 11: 239–45.

Shoji, S., Nanzyo, M. and Dahlgren, R.A. (1993) *Volcanic Ash Soils: Genesis, Properties and Utilization*, Amsterdam: Elsevier.

Simonson, R.W. (1959) 'Outline of a generalised theory of soil genesis', *Proceedings of the Soil Science Society of America* 23: 152–6.

Simonson, R.W. (1978) 'A multiple-process model of soil genesis', in Mahaney, W.C. (ed.) *Quaternary Soils*, Norwich: Geo Books, pp.1–25.

Smettem, K.R.J., Trudgill, S.T. and Pickles, A.M. (1983) 'Nitrate loss in soil drainage waters in relation to by-passing flow and discharge on an arable site', *Journal of Soil Science* 34: 499–509.

Smith, D.G. (1983) 'Anastomosed fluvial deposits: modern examples from Western Canada', *Special Publication of the International Association of Sedimentologists* 6: 155–68.

Smith, D.R. (1992) 'Salinization in Uzbekistan', *Post-Soviet Geography* 33: 21–33.

Smith, R.M., Twiss, P.C., Kraus, R.K. *et al.* (1970) 'Dust deposition in relation to site, season and climate', *Proceedings of the Soil Science Society of America* 34: 112–17.

Soil Classification Working Group (1991) *Soil Classification: A Taxonomic System for South Africa*, Memoirs on Agricultural Natural Resources of South Africa No. 15, Pretoria.

Soil Survey of Scotland (1984) 'Organization and methods of the 1:250,000 Soil Survey of Scotland', Aberdeen: Macaulay Institute for Soil Research.

Soil Survey Staff (1975) *Soil Taxonomy: A Basic System of Soil Classification for Making and Interpreting Soil Surveys*, Agriculture Handbook No. 436, US Department of Agriculture, Washington, DC.

Soil Survey Staff (1992) *Keys to Soil Taxonomy*, Soil Management Support Services Technical Monograph No. 19, Blacksburg, Va: Pocahontas Press.

Soil Survey Staff (1994) *Keys to Soil Taxonomy*, 6th edn, US Department of Agriculture, Soil Conservation Service.

Sokolovsky, A.N. (1930) 'The nomenclature of the genetic horizons of the soil', in *Proceedings of the Second International Society of Soil Science Conference*, Leningrad–Moscow, 5 July, pp. 153–4.

Sollins, P.H., Homann, P. and Caldwell, B.A. (1996) 'Stabilization and destabilization of soil organic matter: mechanisms and controls', *Geoderma* 74: 65–105.

Spyridakis, D.E., Chesters, G. and Wilde, S.A. (1967) 'Kaolinization of biotite as a result of coniferous and deciduous seedling growth', *Proceedings of the Soil Science Society of America* 31: 203–10.

SSEW (Soil Survey of England and Wales) (1983) 'Soil erosion', in *Rothamsted Experimental Station Report for 1982*, Part 1, RES, Harpenden, pp.242–3.

Stace, H.C.T., Hubble, G.D., Brewer, R., Northcote, K.H., Sleeman, J.R., Mulcahy, M.J. and Hallsworth, E.G. (1968) *A Handbook of Australian Soils*, Glenside, South Australia: Rellim Technical Publications.

Stephen, I. (1963) 'Bauxite weathering at Mount Zamba, Nyasaland', *Clay Mineral Bulletin* 5: 203–10.

Stephens, C.G. (1947) 'Functional systems in pedogenesis', *Transactions of the Royal Society of South Australia* 71: 168–81.

Stephens, C.G. (1953) *A Manual of Austrllian Soils*, Melbourne: CSIRO.

Strahler, A.N. (1952) 'Hyposometric (area–altitude) analysis of erosional topography', *Bulletin of the Geological Society of America* 63: 1117–42.

Sugden, D.E. and John, B.J. (1976) *Glaciers and Landscape: A Geomorphological Approach*, London: Edward Arnold.

Swift, M.J., Heal, O.W. and Anderson, J.M. (1979) *Decomposition in Terrestrial Ecosystems*, Oxford: Basil Blackwell.

Syers, J.K., Jackson, M.J., Berkheiser, V.E. *et al.* (1969) 'Eolian sediment influence pedogenesis during the Quaternary', *Soil Science* 107: 421–7.

Szabolcs, I. (1976) 'Present and potential salt-affected soils', *FAO Soils Bulletin* 31: 9–13.

Tardy, Y. (1993) *Petrologie des laterites et des sols tropicaux*, Paris: Masson.

Tate, K.R. (1992) 'Assessment, based on a climosequence of soils in tussock grasslands, of soil carbon storage and release in response to global warming', *Journal of Soil Science* 43: 697–707.

Taylor, G., Eggleton, R.A., Holzhauer, C.C., Maconachie, L.A., Gordon, M., Brown, M.C. and McQueen, K.G. (1992) 'Cool climates lateritic and bauxitic weathering', *Journal of Geology* 100: 669–77.

Tedrow, J.C.F. (1977) *Soils of the Polar Landscapes*, New Brunswick, NJ: Rutgers University Press.

Thornthwaite, C.W. (1931) 'Climates of North America according to a new classsification', *Geographical Review* 21: 633–55.

Thornton, I. (1991) 'Metal contamination in soils of urban areas', in Bullock, P. and Gregory, P.J. (eds) *Soils in the Urban Environment*, Oxford: Basil Blackwell, pp.47–75.

Tice, K.R., Graham, R.C. and Wood, H.B. (1996) 'Transformations of 2:1 phyllosilicates in 41-year-old soils under oak and pine', *Geoderma* 70: 49–62.

Tonkin, P.J. and Basher, L.R. (1990) 'Soil-stratigraphic techniques in the study of soil and landform evolution across the Southern Alps, New Zealand', *Geomorphology* 3: 547–75.

Tricart, J. (1970) *Geomorphology of Cold Environments* (trans.), New York: Macmillan.

Trudgill, S.T. (1977) *Soil and Vegetation Systems*, Oxford: Oxford University Press.

Ugolini, F.C. and Schlichte, A.K. (1973) 'The effect of Holocene environmental changes on selected western Washington soils', *Soil Science* 116: 218–27.

Valeton, I. (1972) *Bauxites*, Developments in Soil Science 1, New York: Elsevier.

van Cleve, K. (1974) 'Organic matter quality in relation to decomposition', in Holding, A.J. *et al.* (eds) *Soil Organisms and Decomposition in Tundra*, IBP Tundra Biome, pp.311–24.

van Post, L. (1922) 'Sveriges geologiska undersöknings trovinventering och nägra av dess hittills vunna resultat', *Sv. Mosskulturföering, Tidskrift* 1: 1–27.

van Vliet-Lanoe, B. (1985) 'Frost effects in soils', in Boardman, J. (ed.) *Soils and Quaternary Landscape Evolution*, Chichester: J. Wiley & Sons, pp.117–58.

Veneman, P.L.M., Vepraskas, M.J. and Bouma, J. (1976) 'The physical significance of soil mottling in a Wisconsin toposequence', *Geoderma* 15: 103–18.

Walker, P.H. (1966) 'Postglacial environments in relation to landscape and soils on the Cary Drift, Iowa', *Iowa State University Agriculture and Home Economics Experimental Station Research Bulletin* 549: 835–75.

Walker, T.W. and Syers, J.K. (1976) 'The fate of phosphorus during pedogenesis', *Geoderma* 15: 1–19.

Wang, C., McKeague, J.A. and Kodama, H. (1986) 'Pedogenic imogolite and soil environments: a case study of spodosols in Quebec, Canada', *Journal of the Soil Science Society of America* 50: 711–18.

Ward, W.H., Burland, J.B. and Gallois, R.W. (1968) 'The geotechnical assessment of a site at Mundford, Norfolk', *Geotechnique* 18: 399–431.

Watanabe, H. and Ruaysoongnern, S. (1984) 'Cast production by the megascolecid earthworm *Pheretima hupiensis*', *Pedobiologia* 15: 20–8.

Watts, S.H. (1983) 'Weathering processes and products under aerial Arctic conditions: a study from Ellesmere Island, Canada', *Geografiska Annaler* Series A, 65: 85–98.

Wauchope, R.D. (1978) 'The pesticide content of surface water draining from agricultural fields – a review', *Journal of Environmental Quality* 7: 459–72.

Webster, R. (1965) 'A catena of soils on the Northern Rhodesia plateau', *Journal of Soil Science* 16: 31–43.

Webster, R. and Burrough, P.A. (1974) 'Multiple discriminant analysis in soil survey', *Journal of Soil Science* 23: 222–34.

Webster, R. and Oliver, M.A. (1990) *Statistical Methods in Soil and Land Resource Survey*, Oxford: Oxford University Press.

Wells, S.G. and McFadden, L.D. (1987) 'Influence of Late Quaternary climatic changes on geomorphic and pedogenic processes on a desert piedmont, Eastern Mojave Desert, California', *Quaternary Research* 27: 130–46.

Wells, S.G., Dohrenwend, J.C. and McFadden, L.D. *et al.* (1985) 'Late Cenozoic landscape evolution of lava flow surfaces of the Cima volcanic field, Mojave Desert, California', *Bulletin of the Geological Society of America* 96: 1518–29.

Whalley, W.B. and McGreevy, J.P. (1983) 'Weathering', *Progress in Physical Geography* 7: 559–86.

Whipkey, R.Z. (1969) 'Storm runoff from forested catchments by subsurface routes', *International Association of Scientific Hydrology Leningrad Symposium Publication* 85: 773–9.

White, R.E. (1987) *Introduction to the Principles and Practice of Soil Science*, Oxford: Basil Blackwell.

Whitfield, W.A.D. and Furley, P.A. (1971) 'The relationship between soil patterns and slope form in the Ettrick association, south-east Scotland', in Brunsden, D. (ed.) *Slopes, Form and Process*, Institute of British Geographers, Special Publication 3, pp.165–75.

Wilde, S.A. (1946) *Forest Soils and Forest Growth*, Walthur: Chronica Botanica.

Willet, I.R., Noller, B.N. and Beech, T.A. (1994) 'Mobility of radium and heavy metals from uranium mine tailings in acid sulphate soils', *Australian Journal of Soil Research* 32: 335–55.

Williams, R.B.G. and Robinson, D.A. (1981) 'Weathering of a sandstone by the combined action of frost and salt', *Earth Surface Processes and Landforms* 6: 1–9.

Wilson, M.J. (1969) 'A gibbsite soil derived from the weathering of an ultrabasic rock on the island of Rhum', *Scottish Journal of Geology* 5: 81–9.

Wilson, M.J. (1986) 'Mineral weathering processes in podzolic soils on granitic parent materials and their implications for surface water acidification', *Journal of the Geological Society of London* 143: 691–7.

Wilson, M.J., Bain, D.C. and McHardy, W.J. (1971) 'Clay mineral formation in deeply weathered boulder conglomerate in north-east Scotland', *Clays and Clay Minerals* 19: 245–52.

Winkler, E.M. and Wilhelm, E.J. (1970) 'Salt burst by hydration pressures in architectural stone in urban atmosphere', *Bulletin of the Geological Society of America* 81: 567–72.

Yaalon, D.H. and Ganor, E. (1966) 'The climatic factor of wind erodibility and dust blowing in Israel', *Israel Journal of Earth Sciences* 15: 27–32.

Yair, A. and Rutin, J. (1981) 'Some aspects of regional variation in the amount of available sediment produced by isopods and porcupines, northern Negev, Israel', *Earth Surface Processes and Landforms* 6: 221–34.

Young, R.A. and Mutchler, C.K. (1969) 'Effect of slope shape on erosion and runoff', *Transactions of the American Society of Agricultural Engineers* 12: 231–3.

Zingg, A.W. (1940) 'Degree and length of slope as it effects soil loss in runoff', *Agricultural Engineering* 21: 59–64.

Zinke, P.J. (1962) 'The pattern of individual forest trees on soil properties', *Ecology* 43: 130–3.

INDEX

abrasion pH, of minerals 53
acidification 53, 82, 192
acidity, of soil 43, 67, 112, 142; *see also* pH
acids: acetic 59; amino 40; carbonic 51; fulvic 66, 68, 69; humic 66; organic 40, 41, 43, 82; sulphuric 136
acrisols 130–1
actinomycetes 15
aeolian input 45, 47, 48, 70, 84, 95, 159
aeolian transport 77
aeration 56
afforestation, affects on soil acidity 192
Afghanistan 151
Africa 62, 131, 136, 148, 150, 155–8, 169, 180
aggregates 22, 31, 113, 186–9; stability 186–7, 199
agriculture, effects on soils 180, 188
albedo: soil 36, 39; rock 50
albite 53
alfisols 79, 111, 154
algae 15
alkalinity, of soil 43, 171
allophane 28, 68, 131
aluminium 53, 66, 67, 68, 77, 83, 96, 98, 112, 113, 145, 192, 193
alluvium 6, 12, 47, 88, 89, 98, 139, 146, 149, 161, 169, 173–4
ammonium 14, 192
amphiboles 28, 55, 86
anatase 80
Andes 131
andisols 111
andosols 85, 131
anhydrite 52
anions 17, 41; *see also individual anions*
annelids 14
anorthite 27
Antarctica 95, 131
anthroposols 120
ants, action in soils 14, 62
Appalachians 159–60
Aral Sea, salinisation of soils 193

arenosols 131
Argentina 112, 148, 194
argon 5
aridification, of soils 199
aridisols 111, 154, 158
aridosols 79, 98
arsenic 196
arthropods 14
atmosphere 5
Atterberg limits 35
attrition 3
augite 86
Australia 62, 80, 111, 113, 120, 131, 134, 136, 148, 150, 151, 158–9, 194

bacteria 15, 58, 59, 192
Barbados 81
basalt 6, 48, 50, 80, 85, 86, 149
base saturation 42, 95, 98, 111, 113, 131, 136, 137, 140, 143, 146; percentage 83
bees, affect on soils 62
beetles, affect on soils 14, 57, 62
beryl 28
bioclimatic zones 155
biomass 9, 14, 83
biosphere 5
biota 82–4
bioturbation 62, 82
birds, affect on soils 63
Borneo 142
boron 194, 196
Brazil 111, 113, 135, 142
British Isles 57, 65, 138, 144, 160–2, 174–5, 182
brown earths 66
brown soils 116
brucite 28
brunisolic soils 119
buffering, of soil 42, 192
bulk density, of soil 29–30, 38, 65, 85, 111, 113, 150, 198

Burkina Faso 82

caesium, in soils as a result of Chernobyl accident 197
calcarosols 120
calcification 67
calcite 6
calcium 82, 111, 112, 135, 138, 140
calcium carbonate 39, 67, 95, 104, 133, 146, 150
calcrete 34, 69
caliche 112
cambisols 132–3, 154
Canada 112, 137, 144, 148
cadmium 196
carbohydrates 16, 40, 59
carbon 14, 40, 83
carbonate 9, 24, 43, 53, 77, 95; accumulation 45, 47,
    111, 138, 149
carbon dioxide 5, 9, 14, 16, 43, 51, 53, 57
carbon monoxide 9
carbon–nitrogen ratio 41, 57, 82, 83, 98, 130, 134, 137,
    138, 139, 140, 142, 145, 146, 148, 150
catenas 80, 165–72, 174
cation, bridging 31; exchange 41–3; exchange capacity
    13, 41, 42, 67, 84, 96, 111, 112, 131, 135, 137, 140,
    142, 196, 197
cations: 17, 196; aluminium 26, 31, 41; calcium 27, 31,
    41, 42, 43, 138, 148; magnesium 31, 42, 43, 148,
    159; potassium 26, 27, 42, 138; sodium 27, 41, 43,
    138, 148
cellulose 40, 56, 57
centipedes, affect in soils 57, 62
chalk 50, 86, 88, 146, 160
chelation 54, 83, 86
cheluviation 66, 67
chernozems 67, 95, 112, 119, 133
Chile 148
China 111
chitin 56
chlorite 29, 80, 81
chromium 196
chromosols 120
chronofunctions 97
chronosequences 96, 98
cicadas, affect in soils 62
clay: accumulation 81, 95; coatings 33, 34, 66, 79, 130,
    137, 148; content 122, 130, 133, 148, 186;
    enrichment index 95; minerals, 6, 23, 28, 68, 76, 77,
    80, 81; translocation 66, 130
climate: clay mineral relationships 80; soil relationships
    75–82, 142, 152–9
coefficient of linear extensibility 36, 113
colour: of soil 39–40, 81, 95, 101, 121, 131, 134, 166,
    167, 171; indices 40, 95–6
colloids 31, 41, 51, 67, 68, 79, 196

consistency, of soil 34
copper 196
cordierite 28
covalent bonding 27
crayfish 63
crickets, affect in soils 62
crotovinas 82, 133
cryosolic soils 119
cryoturbation 36, 63, 120, 137
cyanobacteria 15
cycles: carbon 9–10; hydrologic 8–9; nitrogen, 9, 11, 12

debrisphere 7
decalcification 67, 133
decomposition, oxidative of humus 13
denitrification 12, 191, 194
denudational balance 4
deposition, soil 47, 92
deposits: glacial 12, 95, 96, 137, 139, 174–6;
    solifluction 12; windblown 12
depth, of soil 40, 90, 98, 163
dermosols 120
desiccation, of soils 65, 199
desserts 77, 98, 131, 150, 151, 154, 171
determinative factors, of soil formation 74, 154
diffusion, of air 16; of water 191
diorite 86
dolerite 6
dolomite 6, 50, 51, 53, 86
drainage basin, soil patterns 163–5
drainage, of soil 30–1, 33, 38
drumlins, soil patterns 174
dunes, sand 12, 111, 112, 131, 171
duricrusts 34, 69
duripans 111
dust bowl, North American 180–1

earthworms: activity 14, 62, 79, 83, 112, 133, 136;
    casts 14, 146
ecosystems 5–6
edaphosphere 7
Egypt 194
electrons 54
electrical conductivity of soil water 148, 194
electrostatic forces 31
eluviation 17–18, 65, 66–7, 77, 105
entisols 98, 111, 158
epidote 28
erodibility, of soil 183
erosion 47, 49, 180–9; in British Isles 182–3; in India
    182; by wind 76, 179, 180–1, 188, 189; water 90, 91,
    92, 178, 183–9
eskers, soils of 174
ethane 59

ethylene 59
exchangeable bases 43, 84, 98, 135; sodium percentage 65, 150, 194, 195
eucalyptus scrub, influence on soils 159
evaporation 43
evapotranspiration 8, 75; relationships to soil orders 79
exosphere 5

fabric, of soil 34–5, 62
faecal material 14, 41
feldspars 24, 27, 53, 54, 85, 86, 134, 149; orthoclase 27; oligoclase 53; see also individual minerals
fermentation layer 41
ferrasols 131, 133, 154
ferricrete 34, 69
ferrihydrite 29
ferrosols 120
ferrous sulphide, in waterlogged soils 59, 60
fertilisers 30, 190–2
field capacity, of water in soils 38
Finland 138
floodplains, soils of 47, 89, 112, 173, 174, 193
fluorine 196
fluvisols 136
fragipans 34, 111, 145, 160
France 144
frostheave 79, 171
functional factorial approach 3
fungi 14, 15, 57

gabbro 86
garnet 28
gelifluction 137; see also solifluction
geospheres 5–8
Germany 138
Ghana 142
gibbsite 28, 29, 42, 54, 79, 80, 85, 135
gilgai 52, 113, 150
glacial moraines, soils of 174
gleisation 59
gleying 39, 59, 161, 174, 175
gley soils 116
gleysols 136, 142, 154
Global Soils Degradation Data Base 177–80
global warming 9
goethite 29, 39, 52, 80, 167
gophers, bioturbation effects 82
granite 6, 50, 85, 86, 87, 131, 154, 161
grasshoppers, affect in soils 62
greyzems 137
groundwater 7, 193
gullies 184, 188
gypcrete 34
gypsum 23, 52, 77, 138, 150, 195

haematite 29, 39, 52, 80
halloysite 28, 54, 80
hard setting 23, 199
Hawaii 131
histosols 98, 112, 137, 154
horizons: 1, 94, 101–6, 108, 109, 123; A- 41, 83, 93, 95, 96, 104, 120, 131; argillic 111, 113, 116, 130, 139, 145; B- 18, 95, 98, 131; boundaries 102; C- 95, 96; diagnostic 106, 108–9; eluvial 96, 113, 141, 142, 145; gypsic 148; H- 104; illuvial 120; O- 104; petrocalcic 112
Hong Kong 87
hornblende 86
humification 130, 131, 133, 134, 138
humus: 40, 79, 186; moder 41; mor 41, 82, 113; mull 41, 111, 112
hydraulic conductivity, of soils 44, 194
hydrogen 7, 40, 41
hydrolysis 82, 131, 133, 134, 158; see also weathering
hydrosphere 7

Iceland 13, 48
illite 29, 42, 54, 80
illuviation 18, 67–70, 105, 111
inceptisols 112, 158
India 62, 69, 111, 113, 136, 150, 182
initiating factors, of soil formation 74
inselbergs, soil catenas of 171
involutions 63; see also cryoturbation
inosilicates 26, 27
ionosphere 5
ions, 86, 93, 146; chloride 7; hydrogen 43; hydroxyl 85; mobility 86; silica 27; sodium 7, 66
Iran 151
Iraq 194
iron 54, 66, 67, 77, 83, 96, 98, 112, 113, 134, 166; accumulation with time 81; pan 116, 144, 145, 161
ironstone 169
irrigation 44, 193, 194
isomorphic substitution 27
isopods, affect on soils 63
Italy 180

Japan 131

kandosols 122
kaolinite 28, 42, 54, 70, 79, 80, 81, 85, 112, 113, 134, 143, 145, 148
kastanozems 138, 154
kurosols 122

lakes 174, 193
landscape systems 19
laterite 69, 80

lateritisation 69
leaching 17, 39, 42, 66, 80, 86, 93, 95, 112, 131, 133, 142, 191, 196; index 76, 79
lead, in soils 196
lepdocrocite 29, 59
lessivage 66
levees, soils of 89, 173
lignins 40, 56, 57, 82
limestone 13, 84, 86, 146, 149, 160
lipids 56
liquid limit, of soils 35
lithosols 116, 171
lithosphere 6–7
litter 14, 15, 56, 82, 139, 192; input rates 83
loess 12, 91, 92, 112, 131, 133, 139, 146, 148, 163
luvisolic soils 120
luvisols 66, 139, 154, 158

macrofauna 14
macropores, in soil 30, 64, 65, 191, 196
Madagascar 111
maghemite 29
magnanocrete 34
magnesium 96, 111, 122, 135
magnetite 29
Malaysia 142, 144
manganese 54
mass flow, of soil air 16
mass movement, of soils 90, 91, 92, 168
matric suction 37
mercury 196
mesofauna 14
mesosphere 5
metals, in soils 196; see also individual metals
methane, in soils 59
micas 6, 28, 29, 53, 83, 85, 86, 135, 140, 145
micelles 41
microbial activity 59
microcline 27, 85
micromorphology, of soil 33
millipedes, effects in soils 14, 57
mineralogy, of soils 26–9
minerals, proportion of earth's surface 6
moisture, soil 36–9, 56, 80, 166
mollisols 79, 98, 112, 154
molluscs, in soil 14
molybdenum 196
Mongolia 112
montmorillonite 29, 42, 54, 80, 81, 140, 149
mottling, in soils 38, 39, 59, 78, 121, 135, 137, 142, 144, 146, 148
mudstone 50
Munsell Soil Colour Notation 39
muskrat, affect on soils 82

nematodes, in soils 14
nesosilicates 28
New Guinea 80
New Zealand 47, 76, 80, 84, 96, 131, 132, 142, 162–3
nickel, in soils 196
Nigeria 169
nine-unit land-surface model 167–70
nitosols 139
nitrification 59
nitrogen 5, 16, 40, 56, 57, 59, 82, 83, 98, 190
Norway 96
nutrients 111, 112, 113, 173; depletion 190; of vegetation 82–8

olivine 28, 55
organic matter 9, 13, 30, 39, 40–1, 59, 77, 78, 82, 93, 104, 122, 133, 137–8, 139, 174; decomposition 55–8, 77, 120, 186
organometallic complexes 67, 82
organosols 122
orthosilicates 26, 27, 28
osmotic potential 194
oxidation 54, 121, 147
oxides: aluminium 42, 77; iron 42, 77
oxisols 112, 154, 158
oxygen 5, 7, 16, 40; consumption 16

Pakistan 148, 151
palaeoclimates 95
palaeosols 131, 138
pans 34, 58; plough 30, 63, 198
parent material, of soils 12–13
particles, coatings 24; density 30; mineral 12–13; shape 26
patterned ground 34, 63, 120; see also involutions, permafrost
peat 48, 58, 116, 120, 161, 199
pediments, soils on 171
pedoderm 8
pedogenesis, models of 45–7
pedogenic gradient, of tundra soils 79
pedologic organisation 2
pedomorphic surface 8
pedon 19
pedosphere 7
pedoturbation 61–3, 78, 82
peds 33, 130, 133
periglacial activity 165; see also patterned ground, cryoturbation
permafrost 78, 137, 145
permeability, soil 17, 19, 30–1, 38, 47, 173
pesticides, in soils 30, 195–6
pH 14, 57, 60, 68, 86, 92, 98, 136, 143, 192, 196; see also acidity

phaeozems 139–41
phenols 57, 82
phosphate 14, 41
photophages 55
photosynthesis 14
phyllosilicates 26, 27, 83
phytoplankton 9
piping, in soils 65–6
planosols 141
plastic limit, of soils 36
plasticity index 36
plate tectonics, influence on soils 154, 156
plinthite 34, 134, 158
ploughing, effects on soils 30, 63
podosols 122
podzol 66, 68, 113, 116, 120, 142–5, 154, 161, 193
podzolisation 67, 68, 84
podzoluvisols 145, 154
polypedon 19
polyphenols 66
pores 16, 17, 19, 24, 30, 64, 65
porosity 30–1, 38, 131, 198
Portugal 180
potassium 57, 82, 83, 96, 135, 190
precipitation 75; see also rainfall
principal components analysis, use in soil classification 124
process flux approach 3
profile 2, 46, 47, 98; development index 81, 98
propane 59
propylene 59
proteins 40, 56
protozoa, in soils 14, 15, 57
pyrite, in soils 60, 136
pyroxenes 28, 55, 86

quartz 6, 27, 54, 55

radionuclides, in soils 197
rainfall, 49, 76, 183; acid 53; erosivity 183, 185–7; interception 8
rainforests, tropical 13, 171
rainsplash 35, 76, 183, 185, 198
rankers 116, 145
rats, blind mole, affect on soils 82; pack 82
redox potential 54, 59, 86
reduction 54, 59, 121, 167
regolith 1
regosolic soils 120
regosols 116, 145–6
rendzinas 116, 146
resins 40
rills 91, 184, 188
river terraces, soils of 96, 98, 131

rock, coefficient of volume expansion 50; joints 50;
    mineralogy 50; permeability 50; porosity 50;
    saturation coefficient of 50; specific heat capacity 50;
    tensile strength 50; thermal conductivity of 50; water
    absorption 50
rocks: classification of 6–7; igneous 6; metamorphic 6;
    proportion of earth's surface 6; pyroclastic 6;
    sedimentary 6; volcanic 6; see also individual rock types
roots 42, 51
rubification 40, 95, 98
rudosols 122
Russia 112, 133, 138

salinisation, of soils 111, 193–5, 198
salinity, of soil 43–4, 171
sandstone 13, 50, 86, 160, 165, 169
saprolite 1, 34, 87
saprophages 55
schists 50
scorpions, effects in soils 62
scree, soils of 163
shales 50, 86, 88, 160, 165
silcrete 34, 69
silica 54
silicates 24, 26–8, 86
slaking, of soil aggregates 188, 199
slickensides, effect of swelling clays on 113, 150
smectite 29, 54, 65, 80, 113, 148, 194
snow, effect on soils 163
sodicity, of soil 43–4
sodification, of soils 111, 193–5
sodium absorption ratio 194, 195
sodosols 122
soil, aerobic conditions 16, 54; air, 16–17; anaerobic
    conditions 39, 54, 136, 137; compaction 198; creep
    53, 63, 90, 168; crusting 198; definition 1–3;
    geomorphology 2, 165–76; plasma 3, 33, 34; ripening
    60–1, 136; scaling 199; skeleton 3, 33; waterlogging
    58, 111, 136, 146, 161, 173, 174, 194, 199
soils, arctic 34, 56, 78, 112, 171, 172; alpine 34; azonal
    70, 107; intrazonal 70, 107; marsh 60–1; organic 119,
    199; water 17–19; zonal 70, 107
soilscape 2
soil classification 105–29; Australian 120–2; British
    116–19; Canadian 119–20; FAO/Unesco 113–15,
    130–51; FitzPatrick 122–4; numerical 124–9; Russian
    106–7; U.S. Soil Taxonomy System 107–13
soil loss, equations 91, 184–5
solifluction 161; see also gelifluction, periglacial activity
Solomon Islands 144
solonchaks 146–8, 159, 171
solonetz 120, 142, 148–9, 159
soloth 159
solubility, related to pH 68

solum 2, 34
sorosilicates 28
South America 131, 136; *see also individual countries*
Soviet Union 133
Spain 180
spodosols 79, 84, 108, 113
Sri Lanka 79
state factors 72–3; equations 72–4
steady state, of soil formation 98
stratosphere 5
strontium, in soils 197
structure, soil 31–3, 113, 131, 133, 134, 136, 138, 139, 142, 146, 150, 176, 188, 194
subsidence, of organic soils 199
Sudan 113, 194
sugars, in organic matter 57
sulphate 14, 60, 136; aluminium 136; iron 136
sulphide, ferrous, in marsh soils 136; hydrogen, in waterlogged soils 59, 60, 136
sulphidisation 60, 136
sulphur 54, 60, 136; dioxide 192
Surinam 136
syenite 86
Syria 193
systems, analysis of soils as 3–5, 19–20

tectosilicates 26, 27
temperature, of soil 36, 56, 59
tenosols 122
tephra *see* volcanic ash
terra rossa 39
termites, affect on soils 14, 57, 62, 82, 134
texture 22–5, 132, 169; manual assessment 23
Thailand 62
thermosphere 5
thixotrophic behaviour 34, 79
thresholds, erosional 183; pedologic 45, 81
tidal flushing, of salt marshes 60
titanium, in soils 54
topography, soil relationships 74, 89–93, 163–72, 194
toposphere 8
tourmaline 55
toxins, in soils 194, 196
trampling, affect on soils 30
trees, relationships with soils: coniferous 13, 41, 57, 66, 83, 113, 119, 120, 142, 154, 193; deciduous 41, 57, 66, 83, 120, 132, 139, 154
tropical rainforests, soils of 15
tropics, soils of 42, 54
troposphere 5

ultisols 79, 113, 154, 158, 159
United States of America 47, 62, 65, 79, 81, 83, 95, 111, 112, 131, 137, 138, 148, 150, 159–60, 174
universal soil loss equation (USLE) 184–6
Uruguay 112

vegetation, 95; carr 112; cover 186; fen 112; grassland 41, 83, 112, 133, 139; heath 41, 57, 58, 66, 119, 120, 144; montane 144; moorland, 144; tropical rainforest 77, 136 tundra 119, 141
vermiculite 29, 54, 81, 83, 85, 143, 145
vertisols 52, 98, 113, 149–50, 154, 158, 159
vivianite 59
voids, soil 17, 33
volatilisation 191
volcanic ash 13, 48, 85, 131, 149

wasps, affect on soils 62
water: capillary 18; 36, 37; gravitational 37; hygroscopic 37, 38
water: bypassing 30, 64, 191; depression storage 9; drip flow 8, 66; infiltration 9, 18, 37, 44, 63–4, 82, 90, 188, 194; infiltration capacity 37, 47, 198; infiltration curves 65; infiltration rates 23, 91, 188; overland flow 23, 75, 90, 91, 169, 183, 184, 191, 198; percolation 9, 18, 64–6, 67, 75, 90; soil balance 75, 76; soil shortage 75; soil storage 75; stem flow 8, 84; table 16, 147, 171; throughfall 8; throughflow 9, 37, 64, 75, 90; transmission 186; transmission zone 37
water-holding capacity 37, 76
waxes 40, 57
weathering 47, 49–54, 79, 85–8, 104, 157, 168: biotic 51; carbonation 51–2; exfoliation 50; freeze–thaw 13, 50; hydration 50, 51, 52, 166; hydrolysis 43, 53; mineral sequence 85–6; salt 13, 50, 51; solution 51, 87; thermal 13, 50; wetting and drying 113; unloading 49; weathering; indices 24, 54–5, 105; pits 51; profile 86–8, 134
West Indies 131
wetting front 37
wilting point 38
winter rain acceptance potential (WRAP) 65
woodlice, in soil 14, 63

xerosols 150–1, 154

Zambia 62
zinc, in soils 196
zircon 28, 55, 86
zoophages 55